The Veiled Species of Hebeloma
in the Western United States

Alexander H. Smith is professor emeritus of botany at the University of Michigan.

Vera Stucky Evenson is a mycologist and assistant to the curator at the Denver Botanic Gardens.

Duane H. Mitchel is curator of mycology at the Denver Botanic Gardens.

The Veiled Species of Hebeloma in the Western United States

Alexander H. Smith,

Vera Stucky Evenson, and Duane H. Mitchel

ANN ARBOR   THE UNIVERSITY OF MICHIGAN PRESS

Library of Congress Cataloging in Publication Data

Smith, Alexander Hanchett, 1904–
    The veiled species of Hebeloma in the western
United States.
    Bibliography: p.
    Includes index.
    1. Hebeloma—Classification. 2. Fungi—West (U.S.)—
Classification. I. Evenson, Vera Stucky, 1933–
II. Mitchel, Duane H., 1917–   . III. Title.
QK629.C787S58   1983   589.2′22   83-6644
ISBN 0-472-10036-X

This Publication is Dedicated to Dean Emeritus L. R. Hesler
1888–1977

The University of Tennessee   Knoxville

We dedicate this work to Dean Emeritus L. R. Hesler for his tireless efforts in the study of the Agaricales of the Southeastern United States and in particular for his interest in the neglected genus *Hebeloma*. If it had not been for his interest and persistence, the present contribution would never have been brought to its present stage of completion. Hesler had completed a preliminary manuscript on the genus for North America and was planning to study the Smith collections from Colorado when he was taken ill. The Hesler manuscript will be completed in due time with recent contributions included.

Alexander H. Smith

CONTENTS

INTRODUCTION

Fries (1821) treated *Hebeloma* as a Tribe ("Tribus") of *Agaricus*. The first three words of his description are: "velum marginale floccosum. . . . " *Agaricus fastibilis*, the only species included, he described as follows (as translated): . . . odor nauseous, taste unpleasant; stipe 1.5–3 inches long, subequal, becoming hollow; pileus firm, obtuse, when moist slimy ("viscosus"); lamellae often with serrulate edges bearing droplets; pileus slightly expanded, opaque; stipe squamulose, white; spores subargillaceous. Thus *H. fastibile*, described as a veiled species, became the type of the genus *Hebeloma*. Fries (1838) and subsequent authors have included species with and without veils. In general, the concept of the genus has been followed quite consistently.

As for North America, few comprehensive treatments of the genus are available. Murrill's (1917) treatment is the most extensive but is outdated both as to the number and type of characters emphasized in his classification. Singer (1962 and 1975) placed the genus in the Cortinariaceae between *Inocybe* and *Alnicola*. We emphasize the relationship to *Cortinarius* on the basis of the microanatomy and floristic characteristics. Romagnesi (1965), Bruchet (1970), and Bohus (1972) published works which deal critically with taxa at the species level. We have found, in the central Rocky Mountains of Colorado, some of the species described by them as occurring in central Europe. The evaluation of characters used for establishing species concepts by the above-mentioned authors has been most helpful to us in our study of the species from western North America—particularly those from the central Rocky Mountains. Our study, however, has led to an "explosion" of species in the subgenus *Hebeloma*. This increase in the number of species, as presented here, mirrors the situation previously found in other genera of the Agaricales when the flora of North America is compared quantitatively with that of Europe. As examples we cite *Psathyrella* (Smith 1972), *Lactarius* (Hesler and Smith 1979), *Pholiota* (Smith and Hesler 1968), and *Mycena* (Smith 1947).

Most recent field guides and other works written for the nonspecialist offer little help in either the taxonomy or toxicology of the genus. Smith and Weber (1980) illustrate one species in each subgenus. Miller (1972) illustrates two species. Arora (1979) treats only one veiled species and lists it as poisonous. He states that the presence of a cortina distinguishes it from the common members of the genus. This suggests that veiled species are less "common" than those lacking a veil. The present study, and our preliminary survey of the subgenus *Denudata,* clearly show that both groups contain a large number of species in our western area. Species without a veil are encountered more frequently than the veiled species only because they are typically larger and their fruiting bodies are more conspicuous. However, Arora's observations on the extreme variations in numbers of fruiting bodies from year to year are certainly accurate.

Since Smith is at present finishing the monograph of the North American species of *Hebeloma* by the late L. R. Hesler, which was nearly completed when he was forced to give up the investigation, no extensive introductory discussion is included in the present work other than the description of the genus and details concerned directly in identification of the veiled species. Some general conclusions, however, are suggested as a result of the present study.

The basic taxonomic characters used here to delimit taxa at all levels within the subgenus are those that have long been in use in the classification of the Agaricales. For the most part these are the ones Hesler was using when he studied some of Smith's early collections from Colorado. Hesler considered most of Smith's Colorado collections to belong to undescribed species. It has taken us seven additional collecting seasons to realize that Hesler's first impressions were more right than wrong, and it is with some trepidation that we describe the array of new species in this rather nondescript group of fungi. It is our hope that by the data presented here we can establish some concepts for critical comparison taxonomically and thereby allow some meaningful toxicology to proceed.

The institutions and individuals that have given aid to the general study of *Hebeloma* in North America by advice and loan of specimens, first to L. R. Hesler, then to the present authors, represent a generous cross section of the students of the higher fleshy fungi both in Europe and North America. In this respect the text speaks for itself.

This project over the years, as far as fieldwork is concerned, was carried on as incidental collecting while work was being done on other related projects. Thus the efforts of Mitchel were incidental in a large measure to his collecting of Myxomycetes (Mitchel, Chapman, and Farr 1980). Smith's efforts were incidental to his work on *Mycena* (Smith 1947), financed in part by the Faculty Research Fund of the University of Michigan and incidentally by the National Science Foundation on a series of grants to A. H. Smith for study of western higher fleshy fungi (see Hesler and Smith on *Lactarius* 1979). Each of the present authors has contributed personal financial support to the project, as needed, up to the cost of publication.

Three contributors, however, deserve to be singled out for special mention. First, the team of Virginia Wells and Phyllis Kempton, with assistance from their husbands in the fieldwork, made the best study to date on the mushroom flora of the coastal region of Alaska. Mr. Charles Barrows of Santa Fe, New Mexico, has continued his efforts to collect the fleshy fungi of our semiarid Southwest.

Appreciation is extended to Dr. Sue Pratt of the University of Utah, Department of Biology, for her generosity, her skill with the scanning electron microscope, and her enthusiastic interpretation of the results of the microscopy. The authors also extend thanks to Dr. L. H. Wullstein of that department for his original suggestions and personal support of the project. It should be pointed out that, while these studies have not been used to date in differentiating between species, they have, more importantly, confirmed the homogeneity of the genus.

The Region Covered

The region covered in this work is that area west of the Great Plains north of Mexico to Alaska. It is a large and physiographically diversified region in climate and topography. Collections have been made in the area by Smith from 1935 to the present. Locations in which major collecting effort was centered include the Olympic Mountains of Washington; the northern Cascade Mountains including the Mount Shuksan area; the Mount Hood National Forest of Oregon; the Siskiyou Mountains of southern Oregon; and on the Pacific Coast, from Cape Flattery, Washington, south to Eureka, California. For collections from the part of California south of Eureka we are indebted to Professor Harry D. Thiers of San Francisco State University.

In the Rocky Mountains we have collected in the Priest Lake district of Idaho on several occasions; in the Salmon River district to the south; and, with assistance from Mrs. Ted Trueblood of Nampa, Idaho, collections were obtained from southwestern Idaho and southeastern Oregon. Dr. Kent McKnight contributed collections from the Uinta Mountains of Utah, and Dr. Wilhelm G. Solheim and Smith collected in the Medicine Bow National Forest of Wyoming.

The most intensive collecting effort, however, was made in Colorado where eight successive seasons were spent by various collectors: Shirley W. Chapman, Evenson, Mitchel, and Smith. Contributions of material were made during several seasons by the membership of the Colorado Mycological Society. As a result of this combined effort the *Hebeloma* flora of Colorado had been more intensively studied than that of any other area in North America. Chapman, Evenson, and Mitchel collected extensively in the Front Range from around Idaho Springs to north of Rocky Mountain National Park. Mitchel collected extensively in Eagle County. Evenson, Mitchel, and Smith have collected in eastern Pitkin County in the Savage Lakes area, and around Snowmass Village, as well as in the vicinity of Independence Pass. Many specimens were brought in by those attending the Rocky Mountain Mushroom Conference. The Conference was the focal center for the work in Pitkin County. Barrows collected in southern Colorado and in New Mexico. In Alaska, Kempton and Wells collected extensively as a team. For the most part, their collections will be found in the University of Michigan Herbarium. Their collections represent the largest and best-documented set of specimens that we have seen from our largest state. The specimens forming the documentation of the present study will be found in the Denver Botanic Gardens Herbarium (DBG) and in the Herbarium of the University of Michigan (MICH).

As was evident from the many specimens examined, mixed collections are a problem to those studying veiled *Hebeloma* species. Consequently it did not seem advisable to cite every collection we studied. Also, variation in cheilocystidial morphology in our descriptions is nearly always drawn from a single fruiting body. For further comment on this variation, see pages 6 and 18.

Habitat Relationships

Association with certain species and/or genera of trees has long been known for species of *Hebeloma*. Fries (1838) described the habitat for *H. mesophaeum* as being under pine (". . . Vulgatus in pinetis . . . "). A large number of species are associated with spruce, especially *Picea englemannii* and *P. sitchensis*. Our studies in central Colorado, though preliminary in scope, indicate that as far as habitat relationships are concerned, there is a strong parallel between *Hebeloma* and *Cortinarius* in that there are two relatively distinct floras for each genus: one under hardwood and one under conifers. Thiers's collections from central and southern California have been particularly valuable because they contained many species from hardwood habitats—and much study remains to be done on this particular "flora." The veiled species of *Hebeloma,* though smaller, appear to be most numerous in the conifer habitats. The nonveiled species are more conspicuous only because of their size. Few if any species of either subgenus are truly lignicolous. Those occurring on wood are usually on rotten wood permeated with tree rootlets bearing mycorrhiza. We have not found any species associated with animal remains, but we do have one (*H. amarellum*) which is without question coprophilus.

The physical features of the habitat appear to have some importance in determining a fruiting site. Many species fruit along streams on the outwash areas. These are generally different from those found on seepage areas under herbaceous vegetation on mountain slopes.

Seasonal variations of *Hebeloma* and *Cortinarius* fruitings are also similar. Though not usually part of the "snow-bank" mushroom flora of the Rocky Mountains, some veiled species of *Hebeloma*, e.g., *H. mesophaeum,* and some species of *Cortinarius,* e.g., *C. ahsii,* appear soon after the snow melts. The species found on stream banks after the spring run-off are seldom the same as the ones found on seepage areas in midsummer. July 15 to August 15 has been the major fruiting period for the past eight years in central Colorado for both *Hebeloma* and *Cortinarius.* Usually a week of rainy weather sometime between these dates has triggered heavy fruiting of both genera. In the Pacific Northwest both genera fruit heavily during the fall season: September to October in the Cascades, and along the coast from October through November. *Hebeloma* species at times do fruit in almost unbelievable numbers, as stated by Arora (1979). One can only speculate on the life pattern of the species that show this irregular fruiting pattern. Years of experience collecting in the western area suggest that these abundant fruitings are triggered by ideal conditions of moisture and temperature and an unlimited and readily available food supply. Exactly this same pattern of fruiting has been noted for years here in North America for both genera. Some *Cortinarii* fruit every season given merely adequate moisture and temperature. In about one year in ten, however, tremendous fruitings of "rare" species appear along with those known to fruit regularly. Sudden availability and/or quantity of food seems to have little bearing on this fruiting behavior since *Hebeloma* species are mycorrhiza formers (Hacskaylo and Bruchet 1972).

As far as *Hebeloma* is concerned, species fruit from early spring to late fall in protected plantings as well as in the forest. But over and above all these considerations, one still encounters oddities in the fruiting pattern of the various species. For instance, during the entire summer of 1981 we were able to find hardly a half-dozen collections of veiled *Hebeloma* species, whereas for the previous seven years we collected more than that on almost every day in the field. We would never have embarked on the present project if the 1981 season had been our first.

One of the problems confronting the student of veiled *Hebeloma* species is the problem of mixed collections. These fungi often fruit in a scattered pattern in a restricted area, and the basidiocarps can all closely resemble each other. It is not surprising then, that the toxicology of the genus is almost completely unknown. Given the usual starting point, a case of poisoning, where does one start to determine the culprit? If an experienced taxonomist has difficulty avoiding mixed collections, how can a reliable specimen of the offending mushroom be obtained or an accurate identification made? Without that identification the chemist can do no meaningful work; the toxicologist has no place to start; and the unwary mycophagist and his fellows are no wiser.

Diagnostic Characters

Pleurocystidia. The absence of pleurocystidia has generally been considered a feature of the genus *Hebeloma* (Singer 1975). Pleurocystidia, however, do occur in a relatively small number of species. Hesler (in manuscript) has reported finding them "rare to abundant" in a few species of both the veiled and the nonveiled group. Rarely, in subgenus *Hebeloma*, have we seen these cells in the hymenium near the gill edge and have considered them to be merely a continuation of the cheilocystidia beyond the gill margin.

In the rare instances in which pleurocystidia are found at some distance from the gill margin, we have assumed they are formed there as sterile "scar tissue" cells (cystidia) replacing damaged hymenial reproductive cells (basidia). Parenthetically, we have observed the same phenomenon in *Psilocybe.*

Cheilocystidia. Cheilocystidia are, in our experience, always present in species of *Hebeloma* and are both distinctive and variable (see figs. A11–27). In fact, it is not uncommon to find several morphological types on the edge of a single gill. Differences in shape suggest to us that they originate either from hymenial elements (basidioles) or from extensions of the terminal hyphae of the gill trama. If they arise from tramal hyphae, the basic shape is filamentous. If they arise from basidioles, the basic shape is clavate with modifications.

The filamentous type, in its primitive form, seems to be a simple hyphal prolongation which projects from the gill trama. It is cylindrical to flexuous with essentially parallel walls and never has a basal enlargement. Its apex may be acute, obtuse, or distinctly capitate (see figs. A15, A18). In its most highly developed form it is found in clusters often with the enlarged heads clumped or agglutinated together and covered with slime. The most distinctive aspect of this type of cystidium is that regardless of the shape of the apex, once the walls extending down from the apex become parallel, a ventricose ± basal enlargement does not occur. Therefore, the widest part of the cystidium is above the area in which the walls are parallel. This type is found in both subgenera, but is the predominant type of cystidium in the subgenus *Denudata.*

The clavate type with its variations seems to be, in its most primitive form, a basidiolelike cell that can be differentiated from the basidiole only by its increased size, particularly width. A vesiculose type of cystidium could evolve by further overall inflation. The fusoid-ventricose type (fig. *A*13) apparently arises from the basidiole-type by elongation of the apical region of the cell to produce a ± narrow neck tapered to a more or less obtuse apex. At this stage of development it may be confused with the filamentous type unless one carefully examines the basal region to be sure it still retains some indication of its original enlargement (see fig. *A*21). The distinctive aspect of this type of cystidium, then, is that the widest part of the cell is below the area where the walls become parallel. (Except for the fact that the walls of a circle are parallel for an infinitesimal distance at points directly opposite each other, the above definitions are accurate.) To confuse the issue further, this fusoid-ventricose type of cystidium, contrary to the accepted definition of a cystidium as being an end cell, can form a secondary cross-wall (see fig. *A*24). The differences between this divided cell and other fusoid-ventricose cystidia are: (1) there is no clamp connection at the secondary septum and (2) the proximal portion of the cell is much wider (more ventricose) than the terminal (± extended) portion. The recognition of this unusual cell structure, along with the other variations previously described as the fusoid-ventricose type of cheilocystidium, is very important since this type is indicative of, and found characteristically in the subgenus *Hebeloma*. In some species we find, rarely, a few cheilocystidia in which the apex forks or branches once or twice (fig. *A*26). The significance of this occurrence is not known.

In summary, cheilocystidia in subgenus *Hebeloma* appear to be in an active state of evolution. Since mature basidiocarps occasionally exhibit filamentous, clavate, capitate, and fusoid-ventricose cystidia all on the same gill edge, the cheilocystidium cannot be used to distinguish species as arbitrarily as, for instance, in *Mycena*. We have, however, demonstrated to our own satisfaction that as a rule the presence of a veil, or veil remnants, and the finding of fusoid-ventricose cheilocystidia are reasonably well correlated characters in subgenus *Hebeloma*. As a corollary to this observation, fusoid-ventricose cheilocystidia are even more uncommon to rare in the nonveiled species (subgenus *Denudata*) in which filamentous to filamentous-capitate cheilocystidia are the rule. These latter types of cystidia appear to be clearly related to droplet formation on gill margins. For instance, our study indicates that nearly all species of *Hebeloma* may develop droplets (weep) on the gill margin and apex of the stipe especially when humidity is high. This tendency, however, is very infrequent among veiled species (subgenus *Hebeloma*) in which filamentous-capitate cystidia are few or absent, but is found most consistently in the species of the subgenus *Denudata* which have predominantly filamentous-capitate cystidia which have a tendency to agglutinate. This agglutination process, apparently, is encountered much more frequently in nonveiled species (subgenus *Denudata*). It is correlated with the clumping of capitate cystidia, the concomitant development of slime over the surface of those cystidia, and the resulting visible droplet formation.

Caulocystidia. In the veiled species, we have found caulocystidia to be of little taxonomic value. They resemble the cheilocystidia in size and shape and often exhibit considerable variation on a single stipe. It seems likely that the veil in species of subgenus *Hebeloma* augments the effect of caulocystidia—that of maintaining humidity in the region of the developing basidia.

Pileocystidia. These are not common in subgenus *Hebeloma* although they can occasionally be found as very slender obtuse hyphal end cells in an ixotrichodermium of a young pileus.

The Cuticle of the Pileus. Features of the surface of the pileus are important in the taxonomy of the genus (see figs. *A*28–32). Nomenclature used here for the different patterns of hyphal arrangement has been discussed at length in *North American Species of Hygrophorous* (Hesler and Smith 1963).

The Ixocutis. A simple cutis (without slime) has not been observed by us in subgenus *Hebeloma* (see fig. *A*28). In many species, the hyphae are appressed to the surface and to some extent embedded in slime. The slime may result from the gelatinization of walls of the cuticular hyphae or it may be extruded from the hyphal cells. If the walls of the hyphae are gelatinized, a microscopic examination (in KOH, Melzer's, or water) will show the hyphal walls in unclear focus. On the other hand, if the slime is extruded from the hyphal cells, the hyphae become ± widely spaced but remain sharply defined as observed in the microscope. Both sources of slime can, apparently, be found in a single basidiocarp so the place of origin of the slime is not emphasized in the taxonomy of the subgenus unless it happens to be a prominent and readily discernible feature (see fig. *A*29).

If slime is present in copious quantities, the surface of the pileus will be slimy to the touch when fresh. If little gelatinization or slime-production takes place, the fresh pileus will be only slightly sticky (viscid) to the touch. If the slime has dried in situ, the pileus surface will be "dry" to the touch (not sticky or slimy), but the pileal surface will appear ± shiny as if varnished. Under the microscope, however, the presence of the slime can still be demonstrated on sections mounted in KOH, Melzer's, or in water. In such mounts, if slime is present, the hyphae of the layer are more or less distinctly separated, and the space between them has a translucence not seen in a simple cutis.

The Ixolattice. The ixolattice may be a well-developed ixocutis in which the hyphae have become very widely separated by the slime, and the arrangement presented is modified to that of a tangled mass of hyphae. It is true, however, that at times an ixotrichodermium may collapse and form an ixolattice. By way of standardization of this examination, fresh pilei that are *nearly* mature should be used. In herbarium studies one cannot always be certain on the basis of a single mount which condition prevails.

The Trichodermium and the Ixotrichodermium. As already pointed out, in an ixocutis the hyphae are appressed to the pileus surface. In a trichodermium the elements are ± upright in arrangement and form a turf. This arrangement can originate by the cuticular hyphae giving rise to numerous upright branches, or by hyphae from a layer directly beneath the cutis (often termed a subcutis) elongating and recurving upward. The trichodermial elements may be simple or branched (see fig. *A*30).

If slime is present in the trichodermial layer, the latter is termed an ixotrichodermium. In old pilei, and especially on poorly dried material, the greatly elongated hyphal elements of the ixotrichodermium collapse and rest appressed to the pileal surface. As the slime dries the hyphae become cemented to the pileal surface. Then, when sections of dried material are revived for study, the picture one gets of an ixotrichodermium as seen under the microscope may be that of an ixocutis or ixolattice. Final observations are best made on fresh, nearly mature pilei—as already stated.

Details of Hyphal Cells. The shape of the cells, the arrangement of the hyphae, the diameter of the latter, type of branching or lack of branching, the reaction of the cell walls to chemicals, are all important. Because veil hyphae often remain over the pileus for a considerable time in the development of the basidiocarp, one must be careful not to confuse veil hyphae lying on the cuticle with the cuticular hyphae.

The Hypodermium (or Hypoderm). The layer of hyphae directly beneath the cuticle and distinguishable from the trama by both size and pigmentation is termed the hypodermium. In most species of *Hebeloma* it is visible as a band or zone of brown pigmented hyphae often with associated incrustations of amorphous material on the hyphal walls as revived in KOH or Melzer's reagent. If the hyphal cells in the layer are ± isodiametric to pear-shaped, or in older specimens, inflated but still more or less globose, the hypodermium is termed *cellular* (see fig. *A*31). If the hyphal cells are more or less elongated (cylindric to variously inflated, but still longer than broad) the layer is said to be *hyphoid*, i.e., composed of hyphalike segments. If both cellular and hyphal elements are present in more or less equal ratios, the hypodermium is called *intermediate* (see fig. *A*32). In a pileus, however, most of the component hyphae are to some degree radially oriented. Radial sections of a pileus taken midway between its center and margin will give the truest picture of the degree of "cell differentiation." Sections taken tangential to the pileus will show cross sections of many of the hyphae in the hypodermial zone and groups of these may give a "cellular" appearance since the length of the cell is not shown. Guzman (1980) has indicated that the details of the hypoderm do not serve effectively as a major character in distinguishing genera in the Strophariaceae, especially for separation of *Naematoloma, Stropharia,* and *Psilocybe.* The present study supports Guzman's point of view. In fact, we have used the characters of the hypoderm in *Hebeloma* only where they seemed to have some significance at the species level, as one in a combination of characters, or as an aid in distinguishing stirpes (which often contain a single, or relatively few species). In some species we found that the hypoderm in fresh specimens was not anatomically or chemically distinct from the trama proper, but study of the same material after drying showed an ochraceous to brown hypodermial zone. In a well-developed hypoderm the hyphal walls on revived material are often speckled with heavy dark-colored incrustations. The deep red color observed in the trama of some species in Melzer's (as described herein), has not as yet been evaluated in the systematics of the genus. Here, again, more studies are needed especially on fresh material. This is a difficult evaluation to make since the hyphae of the fresh material mounted in Melzer's remain collapsed for a time, and a color reading taken immediately after the mount is made may fade quickly, as it does on sections of dried specimens in some species.

   The Trama of Pileus and Lamellae. In the lamellar trama the hyphae typically are in parallel to somewhat interwoven arrangements, as seen in cross sections of the gills. The hyphal cells may be short (length-width ratio: 1:1 to 1:3) or longer and cylindric. Old basidiocarps should be used for making observations on size and shape of the hyphal cells. Their walls are essentially thin and colorless in sections of fresh material mounted in KOH. In revived sections the walls are sometimes colored at least as observed on thick sections. The subhymenium usually consists of a very thin layer of branched narrow hyphae and offers little of practical importance in the taxonomy of the group. In a few, however, it is composed of cells being 6–12 μm diam.

   The arrangement of the hyphae in the context of the pileus was found by Hesler to be duplex in some species, and he regarded this as a valuable character. We have found a few species showing it in subgenus *Hebeloma.* It is best observed on fresh specimens fully expanded but not old. Hesler (in manuscript) found it more frequently in the *Denudata.* As currently used, the term *duplex* simply means two-layered (excluding the hypoderm).

   When sections of *Hebeloma* tissue are mounted in Melzer's, different species give color reactions varying from no color change (yellowish to ochre), to a rose-red coloration, a dextrinoid (reddish brown) reaction, to a deep red color. Occasionally "dextrinoid" debris is found scattered in the tramal tissues. We do not regard the rose-red reaction to be a true dextrinoid reaction. At present we have no term to describe this rose-red color; it is not amyloid and it is not dextrinoid. One of its distinctive properties is that of coloring red the mounting fluid as it diffuses through the mount. The deep red color observed in the trama of some species in Melzer's has not yet been evaluated in the systematics of the genus. Here again more studies are needed, especially on fresh material. The above mentioned reactions in Melzer's, while distinct in many cases, have been noted and described but not used taxonomically in this work.

"Chemical Characters." We have found FeSO$_4$ to give some very positive results and have included our data in the descriptions. In most species when a few drops of a 10 percent solution are applied to the base of the stipe, a gray, olive, or (usually) olive-black stain quickly develops. In a small number no stain develops or only a dull olive tone develops in a few minutes. We found one example, however, in which a young basidiocarp gave no reaction, but the older slightly worm-eaten specimens gave an olive-black reaction readily on the damaged areas. The FeSO$_4$ reaction on the pileus was usually negative.

The Spores. The features of the spores are very important in taxonomic studies of species of *Hebeloma*. They fall into four groups: (1) size and shape, (2) the ornamentation, (3) Melzer's reaction of the spore wall, and (4) the color of the spore deposit.

In regard to spore size, the same considerations discussed for *Lactarius* (Hesler and Smith 1979) apply here. Material for measuring spores should be taken from mature pilei. Since in the study of dried material one may not have a deposit available, it is worth pointing out that mounts of spores obtained from crushing bits of gill tissue are likely to show more slightly oversized spores in the mounts than in those taken from spore deposits. The range of spore size in subgenus *Hebeloma* is from 7–17 (19) μm long and (4) 5–8.5 (10) μm in width. Even though differences in spore size are often slight, they are useful in correlation with other taxonomic features. In evaluating spore size, particularly if a deposit shows two rather distinct ranges, one should examine the basidia of that pileus to see if both 2-spored and 4-spored individuals are present. Generally speaking, spores from a single pileus will be larger on 2-spored basidia than on those bearing four spores.

Two main types of *Hebeloma* spores are recognizable with regard to their shape in a profile view of individual spores. In the first type, as shown by *H. mesophaeum* (fig. A1–2, 5) the rounded apex makes the spore somewhat elliptic to bean-shaped in profile. In the second type, the spores appear inequilateral in profile view (fig. A9). The top and bottom lines each angle toward the apex which as a result is blunt to subacute rather than rounded. The term *snoutlike* is used to describe extreme cases of this condition. The terms *elliptic, ovate,* and *inequilateral* apply to two-dimensional views of an object (as seen in a drawing). The terms *ellipsoid* and *ovoid* apply to three-dimensional objects, like eggs. This is important to keep in mind when using the descriptions of spores in this work. For examples of spore shapes, see figures A1–10.

The ornamentation of the spores is important in the identification and arrangement of species of *Hebeloma* as in most other groups of the Agaricales. With the high resolution and magnification obtainable with the SEM electron microscope, the finer ornamentation previously suspected is found to be present on most *Hebeloma* spores (see pls. 10–12). Note, however, that smooth spores do occur in some species (pl. 9). The added detail made evident by the SEM, however, will not be available for the process of making routine identifications of collections. Most mycologists and medical technicians are limited by the range of resolution possible with the light microscope. For this reason we have described spore ornamentation as one observes it under a 3

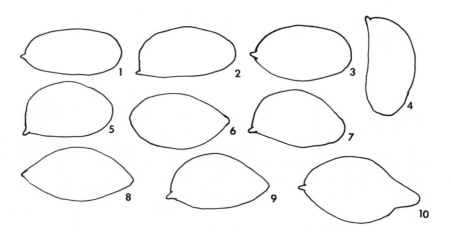

Figs. *A1–10. Terminology of Spore Shapes:* fig. 1, oblong in face view; fig. 2, oblong in profile view; fig. 3, elliptic in face view; fig. 4, bean-shaped in profile view; fig. 5, ± elliptic in profile view; fig. 6, ± ovate in face view; fig. 7, ovate spore in a profile view (± elliptic to obscurely inequilateral); fig. 8, fusiform in face view; fig. 9, inequilateral in profile view; fig. 10, ovate in face view and with snoutlike apex. *Note:* The above terms apply to two-dimensional figures. If one is describing a three-dimensional object the following terms are used for the above figures: fig. 1, cylindric; fig. 2, subcylindric, figs. 3 and 5, ellipsoid, fig. 4, bean-shaped; fig. 6, ovoid (egg-shaped); fig. 7, obscurely inequilateral; fig. 8, fusoid, fig. 9, inequilateral; fig. 10, ovoid but apex snoutlike (in profile such spores are nearly always ± inequilateral in shape).

mm high-dry objective and a 12-× eyepiece. With this optical system those species with very slightly ornamented spores *appear* smooth or practically so. To describe this degree of ornamentation we have often used the term *apparently smooth.* In the final publication on the North American species of *Hebeloma* it is planned to treat spore ornamentation in greater detail.

There are species, however, in which spore ornamentation is obvious under the 3 mm objective, and this degree of ornamentation is

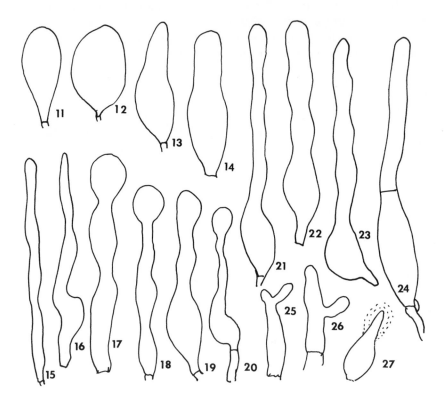

Figs. *A*11–27.   *Terminology of Cheilocystidia:* fig. 11, basidiolelike; fig. 12, saccate; fig. 13, fusoid-ventricose; fig. 14, subutriform (basal enlargement only slightly greater than that of the rounded apex); fig. 15, filamentous; fig. 16, narrowly fusoid-ventricose with elongated neck and apex ± acute; fig. 17, elongate-clavate with flexuous side walls; fig. 18, filamentous-capitate; fig. 19, fusoid-ventricose with ± clavate apex; fig. 20, ± tibiiform (neck narrow, base and apex inflated); fig. 21, fusoid-ventricose with greatly elongated neck and obtuse apex; figs. 22, 23, and 24, fusoid-ventricose with wavy side walls; fig. 24, fusoid-ventricose with a secondary cross-wall distal to the basal inflated portion; fig. 25, apex of cystidium showing a developing bifurcation; fig. 26, a side branch developing on neck of a cheilocystidium; fig. 27, a small (but mature) fusoid-ventricose cystidium with a hyaline "slime-cap."

used in conjunction with other characters to distinguish species. With such spores an oil-immersion objective of NA 2.5, 3, or 4 will be a distinct help. Whether or not spore ornamentation in subgenus *Hebeloma* is the key to the phylogeny of the species appears debatable. But until we have subjected all species to SEM studies, there is little use in defending any one position.

It appears (see pls. 9–12) that the spore has a thin outer layer over it that in the late stages of spore maturation ± disintegrates, or at least is pulled apart (either by the shrinking of the layer itself, or is broken up by the increase in size) as the spore matures. It should be pointed out here that the smooth-spored species (*H. evensoniae*) is atypical in the genus *Hebeloma* both as far as spore ornamentation and the content of the cheilocystidia are concerned.

The dextrinoid character of the spores deserves further comment.

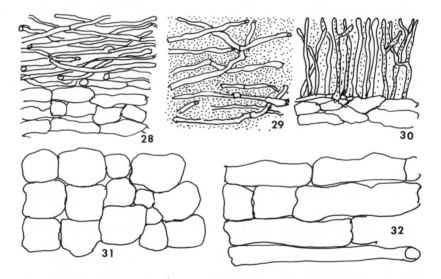

Figs. *A28–32. Terminology of Cuticular and Hypodermial Layers of the Pileus:* fig. 28, a simple *cutis* if slime is not present; fig. 29, the presence of slime (indicated by stippling) makes it an *ixocutis;* fig. 30, a trichodermium or "turf," (if slime is present in the layer it is termed an ixotrichodermium); fig. 31, a short section of a cellular hypodermium (the cells are ± isodiametric); fig. 32, a type of hypodermium intermediate between cellular and hyphoid (some of the component cells are elongated and some ± isodiametric). In a "hyphoid" hypodermium the hyphal cells are predominantly much longer than wide.

The term ixolattice has been used in places in the text. It is a layer of copious slime with cuticular hyphae scattered through it. It may originate in cases where the ixotrichodermium consists of exceptionally long elements and these collapse into a layer as the slime dries. It is to be expected also in old caps featuring an ixocutis in which the hyphae have become widely separated because of the copious slime.

In spore studies with Melzer's reagent in other genera, we have never noted any tendency for the spore pigment to be soluble in the mounting medium. Many species of *Hebeloma*, especially those closely related to *H. mesophaeum*, have spores nearly colorless to yellowish in Melzer's. However, in the present work, we reserve the term *dextrinoid* for a distinctive and reasonably prompt color change of the spores to deep reddish brown. We have found this character useful in distinguishing one species from another. For example, those species with large inequilateral spores tend to have distinctively dextrinoid spore walls that are obviously ornamented. These features appear to be consistent and of value in defining taxa at the various levels below the rank of subgenus. For the present, we consider the red "flush" in the trama when treated with Melzer's, the dextrinoid debris, and the reddish brown spore reaction as independent characters.

The color of the spore deposit has been given prominence in many taxonomic treatments of *Hebeloma* with a subgenus *Porphyrospora* reserved for those few species previously described as having reddish to purple-brown spore deposits. In our work we have found reddish brown spore deposits in both the veiled and nonveiled species, and a gradation of spore color from light clay color to reddish cinnamon to purple-brown without a clear line of separation which could be used taxonomically. Most species in the genus give a deposit which when air-dried is dull cinnamon to clay color to "Verona Brown" (distinctly reddish brown). In such large groups as the Strophariaceae and the Coprinaceae this same spectrum of spore-deposit color is found. In *Hebeloma* species in which the spores appear nearly hyaline under the microscope, the deposit is usually clay color or paler. Mounting such spores in KOH causes an accentuation of the yellow-brown tone, a color change which reminds one of the Strophariaceae, in which the KOH fades the violet component of the spore color and intensifies the yellow component.

Taste and Odor. Appreciation of taste and odor varies markedly from one individual to another. Some people are genetically unable to taste bitter, and the sense of smell is poorly developed and frequently lost in many people. Only fresh material gives sufficient odor to be recognizable as a rule and though some dried specimens have an odor, it is frequently contaminated by insect repellents or other chemical odors in the herbarium. Nevertheless, taste and odor have been used to advantage, in combination with other characters, to distinguish certain species of *Hebeloma*. We have used six general categories:

1. *Sweet* to nauseatingly sweet, usually associated with a fragrant odor, is unusual in the genus but readily recognized by most people.

2. *Bitter,* as in quinine or hops, is occasionally encountered in *Hebeloma.*

3. *Acrid* is often confused with bitter but, as used commonly in mushroom descriptions, applies to the taste of "hot" peppers (a sharply burning sensation).

4. *Radish* (raphanoid) taste and odor is commonly encountered; when strong the odor becomes unpleasantly pungent. In some species, especially if not checked when first collected, the odor is faint and of little value taxonomically. On the other hand, some dried specimens when revived still have a strong raphanoid taste and odor.

5. *Farinaceous* is described only when it is a distinct taste and/or odor. When it is faint or only slightly musty it cannot be distinguished from bland.

6. *Mild* or *bland* as a fungous odor is worth noting, but only as a negative character.

Discoloration of Stipe and Color of Veil. The darkening of the stipe from the base upward has been used as a taxonomic character by students of *Hebeloma* for years. In some species this seems to be a fairly reliable trait, but must be observed on fresh material. Without good field notes, it seems of little value in herbarium studies. The color change represents a chemical process, the expression of which we assume to be a genetically controlled feature. We have given this character less emphasis than did Hesler, but as most previous investigators, we regard it as distinctive within limitations.

The color of the veil is sometimes readily apparent and of taxonomic value. Frequently, however, the scanty, fibrillose veil is noticeable only on very young specimens or can be seen only as patches on the cap or as scattered fibrils on the stipe and therefore its color is difficult to determine. Discoloration of the stipe producing a contrast with the veil color seems to occur with some reliability in a few species. We have not relied on veil color as a distinctive character in this work except when combined with other features.

Concerning Phylogeny

Our study of the veiled species of *Hebeloma* has brought to light some features that we hope will be observed in more detail as *Hebeloma* and related genera are studied further. At present we believe that the genus was derived from *Cortinarius.* Any beginning mycologist will attest to the difficulty of distinguishing these two genera in the field. As we have pointed out, the habitat relationships, seasonal variations in fruiting, and mycorrhizal associations are similar. On this assumption the veiled species of *Hebeloma* would of necessity be the most primitive and closest to *Cortinarius*. We believe the trend in evolution from the veiled to the nonveiled species is paralleled by the trend from clavate and fusoid-ventricose cheilocystidia to the filamentous-capitate and agglutinated cheilocystidia. Many of the smaller species of *Cortinarius* have clavate to fusoid-ventricose cheilocystidia, and it is this type that is characteristic of the veiled species of subgenus *Hebeloma*. In these species the filamentous cheilocystidia are rare and if present are not markedly capitate or clustered, have very little tendency to exude droplets or become agglutinated, and beading of the gill margins is infrequent or inconspicuous. Caulocystidia are also poorly developed in the veiled species and are found mostly at the apex of the stipe. We assume the presence of the veil makes the development of these more highly specialized cystidia unnecessary.

We believe the trend in evolution of the genus is toward the development of cheilocystidia which are basically filamentous with a more or less capitate apex. Some species produce cheilocystidia that gelatinize over the enlarged heads causing the heads to agglutinate into bunches. These in turn, we believe, are the sites of the beads of moisture observed on the gill margins. The same holds for the caulocystidia, which seem to mirror the cheilocystidia in number and form and are the sites at which the beads of moisture appear on the apex of the stipe. The moisture retention, presumably, helps to maintain humidity over the surface of the hymenium. The whole process of cystidial development appears to be functionally oriented to provide moisture (to the sporulating basidia?). In *Hebeloma* this process reaches its maximum development in subgenus *Denudata* as one would expect since in that subgenus there is no veil to serve this purpose. Because of the floristic similarities and the anatomical variations described, we consider the nonveiled species of *Hebeloma* to have been derived from the velate species and these, from *Cortinarius*.

About the Collections

Collection numbers or designations cited throughout this work are those of individual collectors or the official numbers of herbaria such as the herbarium of the Denver Botanic Gardens. Smith's collections cover the period from 1935 to the present. For all collectors, their names or initials are given with their number or whatever designation they used. C. H. Kauffman, for instance, did not keep a record book and did not number his collections. A collection number was made from the date of collecting of the specimen as in "Kauffman 9-7-23." If the collector gave both name and number these of course are so indicated (McKnight 1448 cited as McK-1448 or Wells and Kempton 513 cited W-K-513, etc.). DBG is the official abbreviation of the Denver Botanic Gardens Herbarium and MICH the official designation of the Herbarium of the University of Michigan. The last two mentioned are used mainly to indicate the present location of the collection.

Material Cited. The collections cited are in no way to be assumed to represent all the specimens (or collections) of a given species seen and/or examined in the course of this investigation. They do represent the material, for the most part, from which the descriptions were drawn (both for macroscopic and microscopic characters). We cannot speak for others, but in our study (both collectively and individually) the collections cited were selected from a group of collections all featuring the same or practically the same set of characters. Curiously enough, however, we found that one really good collection will usually serve admirably to characterize a species. The most serious problem with the veiled species of *Hebeloma* involves single basidiocarps. These "singletons" are often found along streams or on the edges of seepage areas, and represent a segment of the flora that a critical investigator cannot afford to overlook. Such single basidiocarps must be studied individually both microscopically and macroscopically—and this usually destroys small fruiting bodies such as those of the *H. mesophaeum* group. With the information obtained from one fruiting body, however, the investigator has a basis for comparing those data with data obtained from other singletons, and usually one finds it possible to decide readily whether he is dealing with a single species or not. If collecting is done carelessly, mixed collections will result, and species concepts will remain confused. As is well known, deviant basidiocarps can be found in clusters of "typical" material of a species (see *H. praecaespitosum*). Such fruitings are well known for *Marasmius oreades* because of the fruiting of the poisonous *Clitocybe dealbata* in clusters of the fairy ring of *M. oreades*.

Collecting and Areas Covered. Collecting in the areas we have enumerated has involved at least three weeks of residence during each of three or more different seasons. But even such coverage over approximately half a century is not sufficient to solve all the problems in the identity of the species one encounters. There is no substitute for resident mycologists with the time and energy to pursue the study of mushrooms in a sizable area. We have tried to cover the situation as it applies to the veiled *Hebeloma* species by using both techniques. To our knowledge, our coverage of the western area is more complete for the veiled *Hebeloma* species than that for any other study yet made on them either in Europe or North America, and yet we regard it as actually preliminary. As mycologists in the western area are aware, climate, and its yearly variations in particular, has an important bearing on which species will fruit and when. Between the variations from season to season, and the sharp difference caused by topography on the general features of the vegetation, which of course are reflected in the local mushroom populations, we as yet have no way of predicting when the "right" season for a particular genus will occur in any given locality. Our knowledge of a given area relative to its mushroom flora is advanced piece by piece by those who can take advantage of the seasons as they occur. The present contribution is an example of this "piece by piece" procedure.

Extralimital Species. The extralimital species from the Great Lakes and northeastern United States are included in the main keys but are designated by a letter within parentheses following the name, for example, *H. affine* (a). These species are numbered at the end of the treatment of the western species as a continuation of the same series of numbers in the text. Hence the appendix is a continuation of the text.

Color Terminology. In the descriptions, color names within quotation marks are taken from R. Ridgway (1912) and were matched with the color sample indicated by the name and under conditions of full daylight. If conditions for accurate color comparison did not prevail, but the Ridgway name is used, quotation marks around the name are not used. "Cinnamon-Brown" means the color was matched under good light conditions. Cinnamon-Brown (without quotation marks) indicates that an actual critical comparison with the color plate was made but is only a close approximation of the color plate of the color involved. In this context, Ridgway names are often used throughout the text, and are the basis of all of Smith's color nomenclature for descriptions of taxa unless indicated otherwise. When not capitalized, the names of the colors are to be interpreted broadly but still approximate the Ridgway terms.

Formulae

The following chemical reagents were used in this investigation in addition to tap water as a mounting medium:

1. Melzer's reagent: KI .... 1.5 g
   I_2 .... 0.5 g
   H_2O .... 20.0 g
   Chloral hydrate .... 22.0 g
   A dark violet to blue or bluish gray color change is termed an amyloid reaction. A dark dull red-brown reaction is designated as dextrinoid. No color change is termed a nonamyloid or inamyloid reaction.
2. KOH (for reviving dried tissues): 2.5 percent aqueous solu-

tion. For demonstrating color changes on fresh material use a 20 percent solution.

3. Alcohol: Use a 70 to 95 percent solution for wetting dried tissue prior to soaking it in water.

4. "FeSO$_4$" (iron salts): Place a crystal or a drop of a ± 10 percent aqueous solution directly on the fungus tissue. The most common reaction is a color change to olive to olive-black.

TAXONOMIC ACCOUNT OF THE VEILED SPECIES OF HEBELOMA IN THE WESTERN UNITED STATES

DESCRIPTION OF
SUBGENUS HEBELOMA

HEBELOMA (Pers. : Fr.) Kummer

Der Führer in die Pilzkunde, p. 80. 1871

Pileus subviscid to viscid or slimy; surface glabrous or with veil remnants variously distributed but usually with at least a zone of fibrils along the margin; color white to ochraceous, grayish, brown or dark reddish brown; odor and/or taste of radish in many species; hypodermium absent to well developed (cellular).

Lamellae broad to narrow (rarely), and adnate to sinuate or rarely decurrent, whitish to brownish or pinkish avellaneous at first, clay color to deep reddish brown at maturity (mostly "Verona Brown" at maturity); edges often beaded with hyaline droplets at first, crenulate to dentate or even.

Stipe (1.5) 2.5–15 (30) mm thick, equal to bulbous, of various lengths depending on the species, solid to hollow; rarely annulate, the veil ± fibrillose and leaving a superior and often evanescent zone where it breaks, remains below this line almost absent to copious and variously dispersed, the fibrils white to gray to ochraceous buff or clay color, rarely with a pinkish tinge.

Spore deposit ochraceous to clay color to various shades of reddish brown or cinnamon brown, rarely purplish brown.

Spores smooth (rarely) to ornamented in some degree, wall about 0.2–0.6 μm thick and typically colored; shape in profile inequilateral to elliptic, ovate to ± bean-shaped, or apex ± snoutlike.

Hymenium.—Basidia typically 4-spored, clavate, rarely with distinctive inclusions such as hyaline globules. Pleurocystidia mostly absent. Cheilocystidia of various types (all on a single gill section in some species), rarely if ever truly thick-walled, in shape clavate, fusoid-ventricose, elongate-clavate, elongate-capitate, or merely filamentous. Caulocystidia when present usually about like the cheilocystidia or longer.

Lamellar and pilear tissues.—Pileal trama hyaline to yellow as revived in Melzer's or in some species bright red to orange but if red then often soon fading. Cuticle of pileus a simple ixocutis to an ixolattice in most species, more rarely an ixotrichodermium; the cuticular

23

hyphae typically 1.2–3 μm diam and usually with clamp connections. Lamellar trama of ± parallel hyphae narrower near the subhymenium than in the central core; subhymenium of narrow filamentous hyphae, rarely containing some small but inflated cells.

Habit, habitat and distribution.—Solitary to gregarious to caespitose on soil in conifer or hardwood forests, on sand dunes, disturbed soil, or on stream banks or outwash areas along streams, spring, summer, and fall; in the Rocky Mountains typically abundant during July and August; in the Pacific Northwest late summer and fall. Most species, apparently, are mycorrhiza formers with forest trees and shrubs, and they occur at all elevations to above the timber line. The genus is present throughout the entire western United States.

The Subgenus
HEBELOMA

The features of this taxon are the presence of a fibrillose veil, often scanty, that when broken usually leaves remnants on the margin of the pileus and/or on the stipe. A second character is the presence of typically fusoid-ventricose, often narrow and greatly elongated cells among those along the edges of the gills (the cheilocystidia). The fusoid-ventricose cystidia are abundant to rare depending on the species being examined.

The type species of the genus is *Hebeloma fastibile* (Pers. : Fr.) Kummer. The subgenus containing the type species of the genus automatically bears the name of the genus.

Key to Major Infrageneric Taxa of HEBELOMA

1. Veil present, usually fibrillose to cortinate .... Subgenus *Hebeloma*
1. Veil absent (but see *H. insigne* which has a "false" veil)
   .... Subgenus *Denudata* (not discussed in this volume)

Key to Sections of Subgenus HEBELOMA

1. Spores in profile view bean-shaped to elliptic or ovate, the apex
   ± rounded (see text figs. *A1–10*) .... Section *Mesophaea* (p. 26)
1. Spores ± inequilateral in profile and narrowed to a blunt apex or
   apex ± snoutlike .... Section *Hebeloma* (p. 84)

Key to Subsections of Section MESOPHAEA

1. Spores 10–15 μm long or longer .... Subsection *Subviscidae* (p. 26)
1. Spores 7–10 (11) μm long .... Subsection *Mesophaeae* (p. 35)

Key to Subsections of Section HEBELOMA

1. Pileus white to pallid or cream color .... Subsection *Pallidae* (p. 84)
1. Pileus more highly colored when young than in above choice 2
   2. When fresh the odor fragrant to pungent-aromatic
      .... Subsection *Praeolidae* (p. 88)
   2. Not as above (the odor if present ± pungent to radishlike) 3
3. Spores 7–10 (11) μm long .... Subsection *Mesosporae* (p. 96)
3. Spores (9) 10–15 μm or longer .... Subsection *Magnisporae* (p. 108)

25

Section MESOPHAEA

Subsection SUBVISCIDAE subsect. nov.

Sporae ellipsoideae vel ovoideae vel in lateralis ad apicem ± obtusae vel rotundatae, 10–15 μm longus.

Typus: *Hebeloma aggregatum*

Spores ellipsoidal in profile with apex ± broadly rounded (not beaked or snoutlike in profile view), and 10–15 μm or more long.

KEY TO SPECIES

1. Habitat on sand dunes, etc.; spores 11–15 × 7–8.5 μm; veil pale gray
.... *H. affine* (a) (p. 162)
1. Not as above 2
   2. Lamellae mottled at maturity, decurrent by a tooth when young;
      spore deposit sepia .... *H. littenii* (g) (p. 169)
   2. Not as above 3
3. Pileus soon radially rimose, obtusely conic to campanulate when young
.... 1. *H. subrimosum*
3. Not as above 4
   4. Odor and taste of radish 5
   4. Odor and taste not raphanoid 6
5. Lamellae pinkish gray; veil pallid .... 2. *H. vinaceogriseum*
5. Lamellae clay color; veil buff .... *H. subfastigiatum* (q) (p. 180)
   6. Odor fragrant; stipe coated with grayish fibrils ... *H. naucorioides* (i) (p. 171)
   6. Odor mild to farinaceous 7
7. Odor and taste farinaceous 8
7. Odor and taste not distinctive 9
   8. Interior of stipe bay-red; hypodermium cellular ..... *H. sterlingii* (o) (p. 178)
   8. Interior of stipe not red; hypodermium hyphoid .... 3. *H. utahense*
9. Cuticle of pileus an ixotrichodermium to an ixolattice 10
9. Cuticle of pileus an ixocutis 11
   10. Stipe fragile, readily splitting lengthwise; growth habit caespitose
.... 4. *H. aggregatum*
   10. Stipe not splitting lengthwise; gregarious .... 5. *H. ollaliense*
11. Spores dextrinoid .... 6. *H. subboreale*
11. Spores not dextrinoid .... 7. *H. wellsiae*

1. Hebeloma subrimosum sp. nov.

Illus. Figs. 1–2.

Pileus 2–4 cm latus, obtusus demum convexus vel campanulatus, glaber, subviscidus, hygrophanus, sordide fulvus vel ad marginem griseobrunneus, demum rimosus; odor nullus; sapor submitis. Lamellae confertae, latae, avellaneae, tarde demum sordide fulvae, ad aciem serrulatae. Stipes 2–3.5 cm longus, 7–10 mm crassus, deorsum brunnescens. Velum pallidum demum argillaceum fibrillosum. Sporae 10–12.5 × 5.5–7 μm, subleves. Basidia tetraspora. Cheilocystidia 37–78 × 3.5–6 × 6–9 μm, filamentosa, ad fusoid-ventricosa.

Specimen typicum in Herb. Univ. Mich. conservatum est, Smith 90148; legit prope Elk Camp, Burnt Mt., Pitkin County, Colorado, 1 Sep 1979.

Pileus 2–4 cm broad, obtuse becoming convex to campanulate or rarely broadly conic; surface subviscid when moist, glabrous in age, margin at first fringed with remains of the pallid veil, surface soon becoming radially rimose; color dull tawny on the disc, the margin a gray-brown ("Wood Brown"), fading to a paler tan (hygrophanous). Context white when faded, firm, both odor and taste mild. KOH on pileus, merely brown; FeSO$_4$ on base of stipe, dark gray.

Lamellae broad, crowded, adnate, avellaneous, becoming ± "Sayal Brown" (a dull cinnamon), edges serrulate under a lens; neither beaded nor spotted.

Stipe 2–3.5 cm long, 7–10 mm thick near apex, base ± enlarged, not readily splitting, when young pallid overall, slowly discoloring brownish in lower portion, apex silky and remaining pallid. Veil fibrillose, pallid but slowly buff colored over area where stipe has darkened, not leaving an annulus or annular zone after breaking.

Spores 10–12.5 × 5.5–7 μm, in profile oblong to ± bean-shaped (rarely slightly inequilateral), in face view ovate to elliptic; surface minutely roughened; not dextrinoid.

Hymenium.—Basidia 4-spored, 8–10 μm broad, clavate, some with highly refractive globules as revived in KOH. Pleurocystidia none. Cheilocystidia 37–78 × 3.5–6 × 6–9 μm, filamentous to narrowly fusoid-ventricose, the necks flexuous and greatly elongated. Subhymenium (revived in Melzer's) reddish fading to orange.

Lamellar and pilear tissues.—Lamellar trama, regular, parallel to slightly interwoven, yellowish in Melzer's (typical of the genus). Cuticle of pileus a very thin ixolattice readily obliterated in sectioning (possibly originating as an ixotrichodermium), the hyphae 3–5 μm diam and only slightly refractive in KOH. Hypodermium of rusty brown-walled hyphae as revived in KOH and radial in arrangement (elements not "cellular"). Some dextrinoid debris noted in mounts in Melzer's but the amount hardly significant. Tramal hyphae ± radially arranged, interwoven, smooth-walled cells of various widths, hyaline to ochraceous, not changing in Melzer's (typical of the genus). Clamp connections present.

Habit, habitat, and distribution.—Scattered under spruce, Burnt

Mt., Pitkin County, Colorado, September 1, 1979, Smith 90148 (type, MICH).

Observations.—The diagnostic features of this species are: (1) the rimose pileus; (2) the pallid veil becoming yellowish; (3) the elongate, narrow, fusoid-ventricose cheilocystidia; (4) the mild odor and taste; and (5) the *Inocybe*-like aspect of the basidiocarps. In most species in which the hymenium is red when first mounted in Melzer's the pileus trama is red also.

2. Hebeloma vinaceogriseum sp. nov.

Pileus 3–5 cm latus, demum plano-umbonatus, subsiccus, tenuiter fibrillosus, violaceo-griseus demum rufobrunneus, glabrescens. Odor et gustus raphaninus. Lamellae "Vinaceous-Buff" demum avellaneae dein argillaceae, confertae, latae, ± adnatae. Stipes 4–6 cm longus, 3–9 mm crassus, aequalis, pallidus, tenuiter fibrillosus, deorsum brunnescens. Velum pallidum. Sporae 9–11 × 5–6 μm, subleves, non dextrinoideae. Basidia tetraspora. Cheilocystidia 44–70 × 5–8 (12) × 4–6 μm, obtusa vel subcapitata.

Specimen typicum in Herb. Univ. Mich. conservatum est, Smith 46587; legit prope Riggins, Idaho, 23 Aug 1954.

Pileus 3–5 cm broad, obtuse, expanding to plane or with a low obtuse umbo, surface dry to the touch (in wet weather), "Benzo Brown" when young becoming "Cinnamon-Drab," and finally reddish brown ("Verona Brown" to "Warm Sepia"), at first thinly coated with fibrils from the veil. Context pallid, odor and taste slightly raphanoid, no color change evident when bruised.

Lamellae "Vinaceous-Buff," becoming avellaneous and finally dingy clay color, close, broad, adnexed, edges uneven.

Stipe 4–6 cm long, 3–9 mm thick, equal, tubular, pallid overall from a thin layer of fibrils and with a thin pallid veil which leaves a slight evanescent zone on breaking, interior pallid above, darkening to dingy brown from the base upward in age.

Spores 9–11 × 5–6 μm, somewhat inequilateral to subelliptic in profile, ovate with a blunt apex in face view, wall fairly thin and appearing smooth, nearly hyaline in water mounts, pallid brownish revived in KOH and pale ochraceous in Melzer's reagent.

Hymenium.—Basidia 4-spored, 26–32 × 6.5–8 μm. Pleurocystidia present only near the gill edge and similar to cheilocystidia. Cheilocystidia 44–70 × 5–8 (12) × 4–6 μm, filamentous with a slightly inflated portion near the base, apex obtuse to subcapitate, thin-walled, smooth.

Lamellar and pilear tissues.—Gill trama typical for the genus.

Cuticle of the pileus an ixocutis of hyphae 3–5 μm diam, clamps present, walls refractive to some extent. Hypodermium hyphoid to intermediate, brown in KOH.

Habit, habitat, and distribution.—Gregarious on wet soil, in a horse corral, Heaven's Gate, Seven Devils Mts., Riggins, Idaho, August 23, 1954 (type, MICH).

Observations.—The pileus is subviscid at best. *H. subviolaceum* is very similar in appearance but its odor and taste are mild, the pileus is blackish brown when moist and mature, and the cheilocystidia are more versiform. The grayish pink tint of the young gills is distinctive for *H. vinaceogriseum* in combination with a grayish veil.

3. Hebeloma utahense sp. nov.

Illus. Figs. 5–6.

Pileus 12–28 mm latus, demum planus, adpresse fibrillosus, alutaceobrunneus; sapor farinaceus. Lamellae latae. Stipes 2–3 cm longus, 2–4 mm crassus, sursum pallidus, deorsum luteobrunneus. Velum subalutaceum, fibrillosum. Sporae 9–12 × 5.5–6.5 μm, in "KOH" leves et subhyalinae, non dextrinoideae. Basidia tetraspora. Cheilocystidia 52–73 × (3) 4–6 μm, ± filamentosa, non agglutinata.

Specimen typicum in Herb. Univ. Mich. conservatum est, McKnight F-1448; legit sub *Pinus contorta*, prope Black's Fork, Utah, 13 Aug 1956.

Pileus 12–28 mm broad, broadly convex to plane, surface matted fibrillose, ± viscid when fresh and moist, "Verona Brown" to "Clay Color"; the margin paler and (in age) pinkish buff. Context thick, pinkish buff; odor distinctive but not identifiable, taste strongly farinaceous.

Lamellae broad, broadly adnexed, close(?), thin, acute at the pileus margin, not beaded.

Stipe 2–3 cm long, 2–4 mm thick, ± equal, brownish pallid above, darker below (as dried), appressed silky or with sparse pale tan fibrils of the veil in a zone near the apex (on dried specimens).

Spores 9–12 × 5.5–6.5 μm, ellipsoid to ovoid (the *H. mesophaeum* type), thin-walled, nearly hyaline varying to brownish in KOH, slightly darker in Melzer's but not dextrinoid.

Hymenium.—Basidia 4-spored, 9–12 μm wide near apex. Pleurocystidia none. Cheilocystidia 52–73 × (3) 4–6 μm diam, subcylindric to filamentous, hyaline, not agglutinated.

Lamellar and pilear tissue.—Lamellar trama typical for the genus. Cuticle of pileus an ixocutis, thin and the hyphae 1–2.5 (3) μm diam, clamped (some hyaline veil hyphae 4–7 μm diam at times matted on

the surface but these not refractive in water or KOH mounts). Hypodermium hyphoid, ± tawny in KOH, speckled with pigment incrustations or wall thickenings or both. Tramal hyphae of pileus merely
yellowish in Melzer's.

Habit, habitat, and distribution.—Scattered to gregarious on soil
under lodgepole pine and willow, August 13, 1956, Black's Fork,
Utah, collected by Kent McKnight (type, MICH).

Observations.—The peculiar odor and farinaceous taste distinguish *H. utahense* from *H. mesophaeum* var. *mesophaeum* and its variants. Also, the veil is apparently pale alutaceous in the young stages.
The lack of a ventricose basal area in most of the cheilocystidia is a
distinguishing microscopic character. The combination of the features
of the cheilocystidia, the hyphoid hypodermium, odor and taste, and
pale brownish spores in KOH amply justify its being recognized at the
species level.

4. Hebeloma aggregatum sp. nov.

Illus. Pl. 5*B;* figs. 3–4.

Pileus 1–3 (3.5) cm latus, obtusus demum convexus vel plano-
umbonatus, ad marginem cinereo-fibrillosus et griseobrunneus, ad centrum cinnamomeo-brunneus, subviscidus; odor et sapor mitis. Lamellae latae, subdistantes, adnatae demum adnexae, avellaneae demum
sordide cinnamomeae. Stipes 4–6 cm longus, 2–3.5 mm crassus, deorsum angustatus dissiliens, deorsum valde fulvus, ad apicem pallidus,
sericeus. Velum cinereo-pallidum demum sordide ochraceum, copiosum. Sporae 9–12 × 6–7 µm, in "KOH" leves, argillaceae. Cheilocystidia versiformes, clavata, fusoide ventricosa vel filamentosa. Cuticula
pileorum gelatinosa.

Specimen typicum in Herb. Univ. Mich. conservatum est, Smith
90057; legit prope Elk Wallow, Fryingpan River, Pitkin County, Colorado, 29 Aug 1979.

Pileus 1–3 (3.5) cm broad, obtuse to convex, often expanded and
umbonate, margin bordered by grayish pallid fibrils from the veil, at
times fibrillose over all except the disc; subviscid to viscid on the disc
(which is glabrous), disc cinnamon-brown, the margin "Wood Brown"
(a brownish gray) but when faded appearing ± canescent and a brownish gray overall. Context white when faded, when fresh brownish gray;
odor and taste not distinctive. FeSO$_4$ staining base of stipe olive-brown,
KOH on pileus cuticle olive-brown.

Lamellae broad, subdistant, adnexed, ventricose, ± avellaneous
becoming ± wood brown (in the brownish gray range), slowly becoming dull cinnamon, edges eroded, not beaded.

Stipe 4–6 cm long, 2–3.5 mm thick, usually narrowed downward, very fragile, readily splitting lengthwise, soon becoming dark rusty brown from the base upward; surface with grayish pallid patches and zoned from the broken veil, these remnants lutescent as stipe darkens, the apex long remaining pallid and silky. $FeSO_4$ on base of stipe dark olive-brown, KOH on pileus cuticle dull olive-brown.

Spores 9–12 × 6–7 μm, in profile obscurely bean-shaped to elliptic or ovate, in face view ovate to oblong; surface slightly punctate in Melzer's reagent, wall not appreciably thickened, clay color in KOH, not dextrinoid.

Hymenium.—Basidia 4-spored, 8–10 μm broad, containing numerous refractive "oil" droplets as revived in KOH. Pleurocystidia none. Cheilocystidia variable: (1) fusoid-ventricose and 32–40 × 7–10 μm, apex obtuse; (2) cylindric to ± clavate and 40–60 × 5–7 μm, apex obtuse; (3) cylindric or filamentous and flexuous, down to a basal portion 7–11 μm broad (but a few ventricose distal to the midportion), not agglutinated.

Lamellar and pilear tissues.—Lamellar trama as revived in Melzer's merely pale yellow but the subhymenium and hymenium orange-red. Cuticle of pileus an ixolattice of tubular hyphae usually having finely roughened walls, the hyphae 3–5 μm diam. Hypodermium hyphoid, dark rusty brown revived in KOH or Melzer's, the hyphae with some wall thickenings and/or incrustations but these not heavy or conspicuous. No dextrinoid debris noted in Melzer's mounts. Pilear trama of ± radial-interwoven hyphae with enlarged cells having smooth walls. Clamp connections present.

Habit, habitat, and distribution.—Caespitose on wet earth under spruce (*P. engelmannii*) above Elk Wallow, Fryingpan River, Pitkin County, Colorado, August 29, 1979, Smith 90057 (type, Mich.)

Observations.—In many species of *Hebeloma* the stipe splits lengthwise fairly readily, but in this species it is difficult to collect specimens without having the stipe splitting into a number of segments. The veil is pallid but, as the stipe darkens, veil remnants in contact with darkened areas slowly become dingy buff. The lack of any distinctive odor or taste also aids in distinguishing this species. It differs from *H. subrimosum* in the clustered habit of the fruiting bodies, the more slender stipe, the nonrimose pileus, the strongly brunnescent stipe, and the pileus being canescent and grayish brown when faded.

5. Hebeloma ollaliense sp. nov.

Pileus 2–3 cm latus, obtusus demum late umbonatus vel convexus, ad marginem fibrillosus demum glaber, ± alutaceus. Odor et gustus mitis. Lamellae adnatae, confertae, angustae demum latae, pallide cinnamomeae. Stipes 2–3 cm longus, 2–4 mm crassus, cavus, deorsum triste alutaceus, sursum pallidior. Velum fibrillosum, pallidum. Sporae 9.5–12.5 (13) × 6–7.5 μm, pallide ochraceae in "KOH," non dextrinoideae, ellipsoideae. Basidia tetraspora, 7–9 μm lata. Cheilocystidia 46–67 × 5–10 × 4–5 × 6–7 μm, plerumque cylindraceosubcapitata ad apicem subcapitata vel deorsum subinflata vel flexuosa filamentosa, saepe agglutinata.

Specimen typicum in Herb. Univ. Mich. conservatum est, Smith 23737; legit prope Ollalie Lake, Mt. Hood National Forest, Oregon, 25 Sep 1946, Gruber and Smith.

Pileus 2–3 cm broad, obtuse with an incurved margin, expanding to broadly convex, plane or plano-umbonate, margin at first with scattered remains of a thin fibrillose veil, soon glabrous, ground color about cinnamon-buff or more argillaceous, only slightly darker when dried, rather evenly colored; odor and taste not distinctive.

Lamellae narrow to finally rather broad, close, adnate, pallid becoming pale cinnamon, either beaded or spotted.

Stipe 2–3 cm long, 2–4 mm thick, equal, hollow, about concolorous with pileus when fresh but base somewhat darker, thinly fibrillous. Veil pallid, not leaving an annular zone on the stipe.

Spores 9.5–12.5 (13) × 6–7.5 μm, appearing smooth but faintly marbled under high magnifications as seen in Melzer's, not dextrinoid (pale ochraceous in either KOH or Melzer's), ellipsoid to ovoid, or some obscurely bean-shaped in profile view.

Hymenium.—Basidia 4-spored, 7–9 μm broad; hymenium dull ochraceous in KOH. Pleurocystidia none. Cheilocystidia abundant, 45–67 × 5–10 × 4–5 × 6–7 μm, cylindric-subcapitate mostly but some slightly enlarged near the base, some flexuous-filamentous, only slightly agglutinated at maturity.

Lamellar and pilear tissues.—Lamellar trama typical for the genus. Cuticle of pileus a collapsing ixotrichodermium of clamped hyphae 1.5–3 μm diam, the hyphal walls refractive. Hypodermium hyphoid, delimited as a rusty ochraceous zone in KOH or in Melzer's, hyphal walls not conspicuously encrusted. Tramal hyphae typical of the genus but compactly interwoven.

Habit, habitat, and distribution.—Gregarious under conifers (mostly *Abies lasiocarpa),* Skyline Road near Ollalie Lake, Mt. Hood National Forest, September 25, 1946, Wm. B. Gruber and A. H. Smith 23737 (type, MICH).

Observations.—This species resembles *H. pascuense* rather closely but is distinguished by having the cuticle of the pileus at first in the form of an ixotrichodermium, in the slightly larger paler spores (in both KOH and Melzer's), and in lacking a raphanoid odor and taste.

6. Hebeloma subboreale sp. nov.

Pileus 1.5–3 cm latus, obtusus vel convexus, demum planus, rufobrunneus ad centrum, ad marginem incarnato-argillaceus, glaber, viscidus, subhygrophanus odor et gustus mitis. Lamellae latae, subdistantes, ventricosae, brunneolae demum cinnamomeae. Stipes 2–4 cm longus, 1.5–2.5 mm crassus, demum triste brunneus. Velum sordide ochraceum. Sporae 10–13 × 6.5–7.5 μm, dextrinoideae, leves in "KOH," ovoideae, apice subacutae. Basidia tetraspora. Cheilocystidia 43–70 × 8–12 μm, elongate fusoid-ventricosa, mox subhyalina.

Specimen typicum in Univ. Mich. Herb. conservatum est, Smith 89006; legit prope Elk Camp, Burnt Mt., Pitkin County, Colorado, 16 Aug 1978.

Pileus 1.5–3 cm broad, obtuse to convex, becoming plane, glabrous, viscid, dark "Verona Brown" over the disc, paler and dull pinkish tan over the margin, opaque at all times, subhygrophanous, when faded tawny around the disc. Context thin, watery, dark brown fading to dingy pinkish buff, $FeSO_4$ staining base of stipe inky black; odor and taste not distinctive.

Lamellae broad, ± subdistant, adnate, becoming ventricose, brownish then pale dull cinnamon ("Sayal Brown"), not beaded and not staining.

Stipe 2–4 cm long, 1.5–2.5 mm thick, equal, at first pale brown and silky, dark brown overall in age. Veil of pale buff fibrils, the remnants forming a thin layer but this soon obliterated.

Spore deposit about "Sayal Brown" either moist or air-dried. Spores (9) 10–13 × 6.5–7.5 μm, pale snuff brown in KOH, slowly dextrinoid, in profile ovate to obscurely inequilateral, in face view broadly ovate, smooth in KOH, faintly marbled as mounted in Melzer's.

Hymenium.—Basidia 4-spored, 24–30 × 7–9 μm. Pleurocystidia none. Cheilocystidia 43–70 × 8–12 μm, fusoid-ventricose and finally greatly elongated, refractive in KOH in some (but the side walls clearly defined and rigid as well as at times faintly brownish and very slightly thickened), some filamentous cells also present, these up to 100 × 5–6 × 6–8 μm, also some cheilocystidia with a secondary septum near the base.

Lamellar and pilear tissues.—Both typical for the genus. Cuticle of pileus a thin ixocutis, the hyphae 2–4 μm diam. Hypodermium cellular, the walls of the cells dark brown in KOH from pigment in the walls and incrustations on them. Tramal hyphae lacking red content in Melzer's mounts. Clamps present.

Habit, habitat, and distribution.—Gregarious on muck under spruce and fir, Elk Camp area, Burnt Mt., Pitkin County, Colorado, August 16, 1978 (type, MICH)

Observations.—This species differs from *H. remyi* in the shape of the cheilocystidia and odor of the crushed flesh. It differs from *H. cistophyllum* Maire in the conspicuously brunnescent stipe and in having a well-developed veil of buff fibrils. From *H. wellsiae* it is distinguished by its dextrinoid spores, in having relatively few filamentous-capitate cheilocystidia, and in the brownish walls of some of the cheilocystidia as revived in KOH.

7. Hebeloma wellsiae sp. nov.

Pileus 10–20 mm latus, convexus demum planus vel leviter depressus, confertae demum subdistantes. Stipes 3–4 cm longus, 2–3 mm crassus, fibrillosus, fragilis, deorsum brunnescens; velum fibrillosum, dilute argillaceum. Sporae 10–13 × 6–7.5 μm, ellipsoideae vel ovoideae, non dextrinoideae. Basidia tetraspora. Cheilocystidia 28–62 × 7–9 μm, elongate-clavata vel cylindraceo-capitata, agglutinata.

Specimen typicum in Herb. Univ. Mich. conservatum est, Wells and Kempton no. 16; legit prope Birch Lake Military Reservation, Fairbanks, Alaska, 22 Aug 1966.

Pileus 10–20 mm diam, convex becoming plane, the disc slightly depressed at times, subviscid, shiny when moisture has escaped; color a medium dull reddish brown (± "Verona Brown") to the ± paler margin. Context concolor with the surface, moist, pallid when faded, about 2 mm thick in the disc; odor and taste not distinctive.

Lamellae moderately broad (2–5 mm), close becoming subdistant, adnate to adnexed, pale tan slowly becoming dingy brown; not beaded.

Stipe 3–4 cm long, 2–3 mm thick at apex, equal, stuffed becoming hollow, pallid, longitudinally appressed fibrillose at first, becoming brown when handled or from the base upward in aging in situ. Veil slight, scanty, remains occurring on pileus margin at first but these soon vanishing, the fibrils pallid tan, no zone or zones noted on the surface of the stipe.

Spores 10–13 × 6–7.5 μm, ellipsoid to ovoid, smooth in KOH,

ochraceous singly and in groups clay color (in KOH), not dextrinoid, wall thin.

Hymenium.—Basidia 4-spored, 8–9 μm broad near apex. Pleurocystidia none. Cheilocystidia 28–62 × 5–6 × 7–9 μm, either elongate-clavate or cylindric-capitate, some slightly ventricose near the base, agglutinated.

Lamellar and pilear tissues.—Lamellar trama typical for the genus or in Melzer's somewhat reddish. Cuticle of pileus an ixocutis but the hyphae 3–5 μm diam, slightly refractive and with fine incrustations or wall-thickenings present; clamps present; walls and hyphal content pale yellow as revived in KOH. Hypodermium hyphoid to cellular and well-developed, somewhat redder in Melzer's. Tramal hyphae of pileus typical for the genus varying to ± reddish in Melzer's (but often soon fading).

Habit, habitat, and distribution.—Gregarious on sandy soil, Birch Lake Military Reservation, Fairbanks, Alaska; among needles and grasses at edge of a bog, August 22, 1966, Wells and Kempton no. 16 (type, MICH).

Observations.—The spore shape is that of *H. mesophaeum* but the size is larger. It differs further in having a pale tan veil and in lacking a distinctive odor and taste. The stipe apparently stains brown rather promptly on injury to the surface if the material is truly fresh. The species is named in honor of the late Mrs. Virginia Wells of Anchorage, Alaska, in recognition of her work on the mushroom flora of Alaska.

Section MESOPHAEA

Subsection MESOPHAEAE subsect. nov.

Sporae 7–10 (11) μm longae.

Typus: *Hebeloma mesophaeum*

Spores 7–10 (11) μm long.

KEY TO STIRPES

1. Cheilocystidia mounted in KOH having a yellow amorphous content when fresh; spores smooth .... Stirps *Evensoniae* (p. 36)
1. Not as above 2
   2. Taste bitter to acrid; spores narrow (3.5–5 μm wide)
   .... Stirps *Chapmanae* (p. 37)
   2. Not as above 3
3. Odor sweetly fragrant; taste mild; veil pale gray ... Stirps *Brunneodiscum* (p. 39)
3. Not as above 4

Stirps EVENSONIAE

Cheilocystidia with yellow amorphous content in KOH when fresh; spores smooth.

8. Hebeloma evensoniae Smith & Mitchel sp. nov.

Illus. Pl. 9.

Pileus (1.5) 2–4 (10) cm latus, obtusus demum convexus vel late umbonatus, viscidus, ad marginem leviter fibrillosus, pallide ochraceus demum argillaceus; odor et gustus raphaninus. Lamellae confertae, adnatae, pallide ochraceae, serrulatae. Stipes 3–8.5 cm longus, 4–8 mm crassus, deorsum attenuatus vel radicatus, brunnescens. Velum copiosum, albofibrillosum. Sporae in cumulis "Vinaceous-Cinnamon," (7.5) 8.5–11.5 × 5.5–6 μm, ovoideae vel ellipsoideae, leves, non dextrinoideae. Cheilocystidia 34–49 × 5–9 μm, fusoid-ventricosa, chrysocystidiis similis. Fibulae adsunt.

Specimen typicum in Herb. Denver Bot. Garden conservatum est, DBG 7948; legit prope "Poorman Gravel Pit," Sunshine, Boulder County, Colorado, 22 May 1978, Vera Evenson.

Pileus 2–4 cm broad, obtuse to convex or broadly umbonate, surface viscid, margin at first with a thin coating of white veil fibrils, pale cinnamon to ± clay color where exposed, disc at times ± ochraceous-tawny. Odor and taste distinctly raphanoid. Context white, unchanging; FeSO$_4$ greenish on stipe base.

Lamellae broad, close, adnate, pale ochraceous becoming grayer and finally dull cinnamon, edges serrulate, not beaded and not spotting.

Stipe 3–8.5 cm long, 4–8 mm thick, tapered to a ± rooting

pointed base, stuffed becoming hollow with a distinct white copious cortina in button stages, brunnescent (± bister) in age, near apex white-pruinose, surface below apex ± appressed white-fibrillose from veil remnants; base with white rhizomorphs.

Spore deposit "Vinaceous-Cinnamon." Spores (7.5) 8.5–11.5 × 5.5–6 μm, ovoid to ellipsoid, apex rounded, smooth in KOH and pale yellow, not dextrinoid, smooth at 7,000 × magnification with SEM.

Hymenium.—Basidia 4-spored, 22–26 × 6–7 μm. Pleurocystidia none. Cheilocystidia 34–49 × 5–9 μm, ± fusoid-ventricose, apex obtuse, internally with clumped refractive material, orange-yellow in KOH when fresh and somewhat resembling that of chrysocystidia.

Lamellar and pilear tissues.—Lamellar trama typical of the genus. Cuticle of pileus a thin ixocutis, the hyphae clamped, yellowish and 3–5 μm diam, encrusted refractive pigment present on the walls. Hypodermium intermediate (both cellular and hyphoid elements present), yellowish in KOH in fresh material. Pileus trama compactly interwoven, yellowish to hyaline in KOH, the cells often greatly inflated.

Habit, habitat, and distribution.—Gregarious under young cottonwoods on litter in an old granite pit, Poorman Gravel Pit, Sunshine, Boulder County, Colorado, May 22, 1979, legit Vera Evenson (type, DBG).

Observations.—The stature, raphanoid odor and taste, and the color of the mature gills indicate the genus *Hebeloma,* but the cheilocystidia are aberrant insofar as their content is concerned, and the color of the spore deposit is not typical of the genus. It is, however within the range we have found for the genus. The pale yellow color of the spores in KOH is a feature not infrequently encountered in other species of the genus, and the absence of ornamentation on the spores as seen in SEM studies (pl. 9) is not typical of the genus. Ricken (1915) described cheilocystidia with a yellow content for *Hebeloma punctatum.* The cheilocystidia of *H. evensoniae,* however, are larger and not threadlike, the stipe base slowly becomes very dark brown, and the spores are not exceptionally thick-walled for the genus.

Stirps CHAPMANAE

Spores narrow, 3.5–5 μm wide; taste bitter.

9. Hebeloma chapmanae sp. nov.

Pileus ± 3 cm latus, plano-convexus, saepe umbonatus, glabrescens, subviscidus, argillaceus vel ochraceus ad marginem, ad centrum

luteobrunneus; gustus amarus, odor raphaninus. Lamellae latae, confertae, argillaceae. Stipes ± 5 cm longus, circa 5 mm crassus, brunnescens, striatus. Velum ochraceum, fibrillosum evanescens. Sporae 7.5–9
× 4.5–5 μm, subleves, ellipsoideae vel ovoideae, non dextrinoideae.
Cheilocystidia 34–60 × 5–7 μm, filamentosa vel ad basin leviter ventricosa. Cuticula pileorum gelatinosa.

Specimen typicum in Herb. Denver Bot. Garden est, Shirley
Chapman (DBG 2473); legit prope Sacramento, Park County, Colorado, 13 Sep 1969.

Pileus ± 3 cm broad, plano-convex but with a small umbo, glabrous (veil remnants buff colored and soon obliterated), surface
slightly viscid; colors tan over the disc and clay color to buff over the
margin. Context thin, tan, watery, odor raphanoid, taste bitter—
lingering and almost acrid.

Lamellae broad, close, sinuate, extending beyond the edge of the
pileus, buff colored, as dried almost clay color.

Stipe ± 5 cm long, about 5 mm thick, slender, equal, darkening
below, longitudinally striate with tan polished striations; remnants of
the thin veil soon evanescent. Spores 7.5–9 (10) × 4.5–5 μm, oblong
to a few ovoid, smooth (apparently) in KOH, ± thin-walled and pale
orange-yellow (in KOH), not dextrinoid, some with a large (oil ?) drop
within.

Hymenium.—Basidia 4-spored, 22–24 × 8–9 μm, clavate, yellowish in KOH. Pleurocystidia none. Cheilocystidia 34–60 × 5–7 μm,
filamentous with apex blunt to slightly enlarged, or ± ventricose near
the base.

Lamellar and pilear tissues.—Lamellar trama typical for the genus. Cuticle of pileus an ixocutis, the hyphae 3–6 μm diam, yellowish
in KOH. Hypodermium "cellular" (from cut ends of hyphae), ± hyphoid in radial sections, dark yellow-brown in KOH, cells up to 15 μm
diam. Pileus trama typical of the genus (the hyphae of the pileus trama
± radially disposed). Clamps present.

Habit, habitat, and distribution.—Caespitose in moss on shaded
black soil, collected at Sacramento (west of Fairplay), Park County,
Colorado, September 13, 1969, by Shirley Chapman (type, DBG); also
collected by Charles Barrows, no. 2021, in New Mexico.

Observations.—The color of the pileus closely resembles that of
H. pascuense but the striations on the stipe readily distinguish it in the
field. We have, to date, seen this feature on only one other species,
and it belongs in subgenus *Denudata*. The narrow spores, persistently
bitter to ± acrid taste, thin veil, striate stipe, and pale pileus are
distinctive as a package.

Stirps BRUNNEODISCUM

Veil pale gray; odor fragrant.

10. Hebeloma brunneodiscum sp. nov.

Pileus 1.5–3.5 cm latus, obtuse conicus demum ± planus, ad marginem griseofibrillosus, glabrescens, obscure rufobrunneus demum spadiceus; odor fragrans, gustus mitis. Lamellae pallidae deinde "Vinaceous-Buff" (griseoincarnatae) demum obscure cinnamomeae, latae, adnatae, secedentes, non maculatae. Stipes 4–7 cm longus, 4–7 mm crassus, deorsum atrobrunneus, sursum pallidus. Velum cinereum, evanescens. Sporae 8–10 × 6–6.5 μm, non dextrinoideae, ellipsoideae vel subphasaeoliformes vel leviter inequilaterales, subleves. Cheilocystidia 42–68 × 7–12 μm, fusoid-ventricosa, collo elongato et flexuoso, ad apicem obtusa.

Specimen typicum in Herb. Univ. Mich. conservatum est, Smith 86872; legit prope Independence Pass, Pitkin County, Colorado, 19 Jul 1976.

Pileus 1.5–3.5 cm broad, obtuse to convex becoming nearly plane, viscid and shiny but with faint patches of grayish fibrils on marginal area, finally glabrescent; color near "Verona Brown" on margin and "Warm Sepia" over the disc, some ± streaked and mottled. Context watery brown, soft, moderately thick, odor fragrant, taste mild. FeSO₄ no reaction.

Lamellae moderately broad, close, depressed-adnate, pallid becoming ± vinaceous-buff and finally Verona brown, horizontal, edges even and concolorous with faces, not spotted.

Stipe 4–7 cm long, 4–7 mm thick, equal, fibrous, soon "Mummy Brown" near base and the change progressing upward, pallid and silky near apex. Veil thin and grayish, all evidence of it on stipe soon vanishing.

Spores 8–10 × 6–6.5 μm, ± smooth, elliptic in face view, in profile slightly inequilateral to slightly bean-shaped or oblong, pale clay color in KOH, not dextrinoid, apex ± rounded.

Hymenium.—Basidia 4-spored, 32–36 × 8–10 μm, clavate. Pleurocystidia none. Cheilocystidia abundant, 42–68 × 7–12 μm, mostly slightly ventricose near base and with a long neck often wavy in outline, apex obtuse, thin-walled.

Lamellar and pilear tissues.—Lamellar trama typical for the genus. Cuticle of pileus an ixocutis of brown hyphae 2–3 μm diam and originating from a hypodermium of inflated cells having rusty brown walls.

Habit, habitat, and distribution.—Gregarious under mixed conifers, Independence Pass, Pitkin County, Colorado, July 19, 1976, Smith 86872 (type, MICH).

Observations.—The cuticle of the pileus may possibly originate as an ixotrichodermium, but if so the elements soon become appressed to the surface of the pileus and as revived in KOH give the impression of an ixocutis. The grayish veil, distinctly fragrant odor, mild taste, lamellae which become pale grayish pink before the spores mature, along with the weak reaction of the base of the stipe to $FeSO_4$ form the distinctive combination of characters.

H. versipelle (Fr.) sensu Romagnesi (1965; p. 322) is very close to our *H. brunneodiscum,* but the base of the stipe in the latter becomes blackish brown, its taste is mild, and its veil is gray. We cannot be sure that the odor of *H. versipelle* as described by Romagnesi (l.c., "d'herbe de persil") is the same as that present in *H. brunneodiscum,* and so are assuming that they are different—at least until more collections can be studied.

Stirps REPANDUM

Spores dextrinoid.

11. Hebeloma repandum Bruchet

Bull. Soc. Linn. de Lyon, 39 année: 50. Jun 1970

Illus. Bruchet, l.c., pl. xii.

Pileus 2–3 cm broad, obtuse with a spreading margin and incurved edge, becoming ± expanded-umbonate, viscid, dingy clay color over disc, marginal area grayer (near avellaneous but tinged with yellow) and usually with a slight zone of appressed veil fibrils that is inconspicuous. Context firm, pallid when faded, odor slight, taste distinctly bitter-pungent (almost acrid); with $FeSO_4$ instantly olive-fuscous on lower part of stipe, the color change slower elsewhere; KOH gives no significant reaction, dextrinoid in Melzer's reagent.

Lamellae narrow, close, adnate, edges uneven, surfaces pallid becoming a dull dingy brown, not beaded.

Stipe 3–4.5 cm long, 3–4 mm thick at apex, hollow, firm, evenly enlarged downward or equal, dingy brown below, whitish near apex, thinly fibrillose from a thin veil and leaving a faint veil line for a time near the apex, merely silky above the veil-line.

Spores 9–11 × 5–6 μm, appearing smooth, pale clay color in KOH, in face view subelliptic to subovate, in profile the ventral line nearly straight and the dorsal line convex, dextrinoid.

Hymenium.—Basidia 4-spored, 30–35 × 7–8 μm. Pleurocystidia none. Cheilocystidia fusoid-ventricose, 35–50 × 7–11 μm, thin-walled, hyaline as revived in KOH.

Lamellar and pilear tissues.—Lamellar trama typical for the genus. Cuticle of pileus an ixolattice, the hyphae 1.5–3 μm diam, yellowish in KOH but soon hyaline, clamps present. Hypodermium cellular in tangential section, somewhat so in radial sections, rusty ochraceous in KOH.

Habit, habitat, and distribution.—Gregarious under brush and conifers, Independence Pass area, Pitkin County, Colorado, July 23, 1976, Smith 86910.

Observations.—The collection described here appears to belong in *H. repandum*. The diagnostic combination of features is the harsh taste, narrow close gills, brunnescent stipe, medium-sized spores, and rather ordinary cheilocystidia for the subgenus.

Stirps PSEUDOSTROPHOSUM

Cuticle of pileus an ixotrichodermium which may collapse to form an ixolattice especially in mature or older specimens.

KEY TO SPECIES

1. Stipe soon yellow-brown overall; veil ochre-yellow; lamellae narrow
   .... 12. *H. luteobrunneum*
1. Not with above combination of characters 2
    2. Stipe ± annulate from a zone of white fibrils; stipe 2–4 mm thick;
       odor pungent .... 13. *H. subannulatum*
    2. Not as above 3
3. Both odor and taste mild; veil thin and pallid; fresh pileus dark date brown
   .... 14. *H. riparium*
3. Not as above 4
    4. Odor and taste of radish; lamellae broad; pileus slowly staining
       blackish brown .... 15. *H. substrophosum*
    4. Not as above 5
5. Pileus deep vinaceous-brown at first; veil white; gills not staining
   .... 16. *H. vinaceoumbrinum*
5. Pileus alutaceous to cinnamon before fading; veil when young buff colored
   6
    6. Cheilocystidia 40–70 × 7–12 × 2.3–5 μm .... 17. *H. pseudostrophosum*
    6. Cheilocystidia 28–43 × 5–7 × 7–9 μm .... 18. *H. alpinicola*

12. Hebeloma luteobrunneum sp. nov.

Pileus ± 3 cm latus, convexus vel subplanus, leviter viscidus, pallide spadiceus, ad marginem fibrillosus. Velum ochraceum, evanescens. Odor et gustus submitis. Lamellae connfertae, angustae, ± cinnamomeae. Stipes 3–5 cm longus, 2–4 mm crassus, aequalis, luteobrunneus, zonatus. Sporae 9–11.5 × 5–6 μm, ellipsoideae vel ovoideae, non dextrinoideae. Pleurocystidia nulla. Cheilocystidia 32–69 × 5–9 μm, fusoid-ventricosa, obtusa. Cuticula pileorum ixotrichoderma est. Hypodermium cellularum.

Specimen typicum in Herb. Univ. Mich. conservatum est, Smith 88819; legit prope Independence Pass, Pitkin County, Colorado, 31 Jul 1978.

Pileus ± 3 cm broad, convex to nearly plane, slightly viscid, ± "Snuff Brown" (a dingy yellow-brown or dull date color), marginal area at first dotted with the buff remains of the veil. Odor and taste not distinctive.

Lamellae narrow, close, adnate, dingy pinkish tan (± cinnamon), not beaded and not spotted.

Stipe 3–5 cm long, 2–4 mm thick, equal, yellow-brown to the apex, zoned with ochraceous-buff remnants of the veil and an annular zone ± persistent.

Spores 9–11.5 × 5–6 μm, ellipsoid to ovoid, minutely roughened, not dextrinoid (merely yellowish in Melzer's).

Hymenium.—Basidia 4-spored, 24–28 × 5–7 μm, projecting when sporulating. Pleurocystidia none. Cheilocystidia numerous, 32–69 × 5–9 μm, fusoid-ventricose, necks elongating, apex obtuse, not agglutinating.

Lamellar and pilear tissues.—Lamellar trama typical for the genus. Cuticle of pileus an ixotrichodermium, the elements narrow and clamped at the septa. Hypodermium cellular, rusty brown in KOH. Pilear trama of hyaline hyphae, the cells ± vesiculose.

Habit, habitat, and distribution.—Scattered under conifers, Independence Pass area, Pitkin County, Colorado, July 31, 1978, Smith 88819 (type, MICH).

Observations.—The diagnostic characters are: the dingy yellow-brown pileus, pale ochre-yellow veil, the narrow gills and the ixotrichodermium of the pileus. The short, inflated cells of the pileal trama are a possible additional character, but we hesitate to emphasize it at present.

13. Hebeloma subannulatum sp. nov.

Pileus 1–2.5 cm latus, obtusus deinde plano-umbonatus, ad marginem albofibrillosus, subviscidus, alutaceus; odor pungens, gustus mitis. Lamellae latae, confertae demum subdistantes, ventricosae, adnatae, obscure cinnamomeae. Stipes ± 3 cm longus, 2–4 mm crassus, deorsum brunnescens, sursum pallidus fibrillosus; annulus albofibrillosus. Sporae 8.5–10 × 5–5.5 μm, subleves, in "KOH" pallide ochraceae, ovoideae vel ellipsoideae. Basidia tetraspora et bispora. Cheilocystidia 38–45 × 5–6 μm, cylindrica vel anguste clavata.

Specimen typicum in Herb. Univ. Mich. conservatum est, Smith 89095; legit in graminis, prope Independence Pass, Pitkin County, Colorado, 21 Aug 1978.

Pileus 1–2.5 cm broad, plane with a slight umbo, surface dry to the touch, thinly white fibrillose overall, ground color alutaceous, in age with appressed patches of veil fibrils along the margin. Context thin, grayish pallid fading to whitish; odor strongly pungent, taste slight; FeSO$_4$ on stipe base olive-fuscous.

Lamellae broad becoming ventricose, close, adnate, finally subdistant, dull cinnamon at maturity; edges not beaded or stained.

Stipe about 3 cm long and 2–4 mm thick, equal, pallid above, becoming brown below, with a whitish fibrillose superior annulus, fibrillose below the annulus. Veil white and copious.

Spores 8.5–10 × 5–5.5 μm, ovoid to ellipsoid, in KOH under high-dry lens ± smooth, not dextrinoid, pale ochraceous in KOH.

Hymenium.—Basidia 4- and 2-spored, 28–30 × 5–7 μm. Pleurocystidia none. Cheilocystidia 38–45 × 5–6 μm, narrowly clavate to cylindric, obtuse, basal area scarcely enlarged.

Lamellar and pilear tissues.—Lamellar trama typical of the genus. Cuticle of pileus an ixotrichodermium, the elements slender (2–3 μm diam), clamped, hyaline. Hypodermium cellular, the cells 10–20 μm diam, with walls heavily encrusted with yellow-brown material. Trama of pileus typical for the genus.

Habit, habitat, and distribution.—Solitary in grass, spruce nearby, Independence Pass area, Pitkin County, Colorado, August 21, 1978, Smith 89095 (type, MICH). It was also collected by Kauffman at Tolland, Colorado, August 26, 1920.

Observations.—This species might possibly be *H. testaceum* of Europe but we did not find cheilocystidia with secondary septa as shown by Bruchet, and neither the odor nor taste "fit" his description well. Moser (1978) gives the spores of *H. testaceum* as (8) 10–12 (13) × 5–6.5 μm and lists the habitat as moist places under hardwoods.

14. Hebeloma riparium sp. nov.

Pileus 1–3.6 cm latus, obtusus demum convexus vel umbonatus, spadiceus, demum obscure cinnamomeus, canescens, odor et gustus mitis. Lamellae "Vinaceous-Buff" demum sordide cinnamomeae, confertae, angustae, adnatae. Stipes 2–3 cm longus, 4–6 mm crassus, pallidus deinde brunnescens, fibrillosus. Velum fibrillosum, ± evanescens, pallidum. Sporae 8–10 × 5–5.5 μm, ± phaseoliformes, non dextrinoideae, subleves. Cheilocystidia 33–46 × 9–12 μm, demum 50–67 × 7–10 μm, fusoid-ventricosa, obtusa.

Specimen typicum in Herb. Univ. Mich. conservatum est, Smith 88621; legit prope Lincoln Creek, Pitkin County, Colorado, 16 Jul 1978.

Pileus 1–3.6 cm broad, obtuse to convex, the margin incurved, becoming plane or nearly so, or margin finally uplifted, ± evenly "Prout's Brown" to "Buckthorn Brown" (dark to medium yellow-brown), opaque at all times, slowly becoming pale "Sayal Brown" in aging, canescent at first and avellaneous over marginal area. Context thin watery brown, fragile, odor and taste mild, $FeSO_4$ staining base of stipe gray (very slowly); KOH on cuticle of pileus no reaction.

Lamellae narrow, close, adnate, pallid becoming a pale vinaceous-buff, near "Sayal Brown" in age, not beaded and not stained.

Stipe 2–3 cm long, 4–6 mm thick, equal, pallid from a thin coating of veil fibrils, becoming bister from the base up, solid, surface fibrillose from the thin veil but no annular zone evident.

Spores 8–10 × 5–5.5 μm, elliptic to ovate in face view, bean-shaped to very obscurely inequilateral in profile, pale snuff brown in KOH and about the same in Melzer's, appearing smooth under a high-dry objective.

Hymenium.—Basidia 4-spored, 6–7 (8) μm broad. Pleurocystidia none. Cheilocystidia fusoid-ventricose, at first 33–46 × 9–12 μm but becoming 50–67 × by 7–10 μm by the elongation of the neck, apex obtuse, not branched or forked, lacking secondary septa.

Lamellar and pilear tissues.—Lamellar trama typical for the genus but hymenium and subhymenium reddish to orange in Melzer's. Cuticle of pileus a tangled ixotrichodermium collapsing to an ixolattice, hyphae 2–3.5 μm diam and with refractive walls, in Melzer's reagent copious dextrinoid debris forming; clamps present. Hypodermium of yellow-brown inflated cells (as seen in KOH mounts of fresh material), tawny as revived in KOH. Tramal hyphae of pileus typical of the genus (no red content in Melzer's).

Habit, habitat, and distribution.—On soil on an overflow area of a stream, Lincoln Creek, Pitkin County, Colorado, spruce nearby, July 16, 1978 (type, MICH).

Observations.—This species illustrates one reason why mature specimens should be used for study of the cheilocystidia. At a certain stage in the development of the fruiting bodies the cystidia apparently elongate rather rapidly. One would expect them to agglutinate in age.

The ixotrichodermium of the pileus, narrow close gills, and absence of a distinctive odor and taste distinguish this species.

15. Hebeloma substrophosum sp. nov.

Illus. Pl. 3; figs. 17–18.

Pileus 3–7 cm latus, obtusus demum convexus vel subplanus, sordide cinnamomeus, ad marginem subsquamulosus, ad centrum sordide vinaceobrunneus deinde atromaculatus, subviscidus; odor et gustus raphaninus. Lamellae adnatae, brunneae, fuscomaculatae confertae, latae. Stipes 6–10 cm longus, 8–13 mm crassus, deorsum demum atrobrunneus, fibrillosus; velum subargillaceum. Sporae 9–12 × 5.5–6.5 μm, obscure inequilaterales, non dextrinoideae. Cheilocystidia 30–65 × 7–9 μm anguste clavata vel deorsum subventricosa. Cuticula pileorum ixotrichoderma est.

Specimen typicum in Herb. Univ. Mich. conservatum est, Smith 87043; legit prope Independence Pass, Pitkin County, Colorado, 4 Aug 1976.

Pileus 3–7 cm broad, obtuse to convex, expanding to plane or nearly so; disc glabrous dull cinnamon and subviscid; margin pallid from a dense coating of fibrils or patches of them, in age the disc more or less "Verona Brown" but with blackened areas. Context dingy pale pinkish buff, thin but firm at first; odor and taste raphanoid; $FeSO_4$ greenish gray on stipe; KOH brownish on surface of pileus.

Lamellae broad, close, adnate, pale dull brown becoming dingy "Verona Brown," stained darker in age.

Stipe 6–10 cm long, 8–13 mm thick, pallid above, dark brown in the base and progressively slowly upward; equal or narrowed at the base. Veil fibrillose remains soon evanescent, in color pallid to dingy buff.

Spores 9–12 × 5.5–6.5 μm, subelliptic to very obscurely inequilateral in profile and some obscurely bean-shaped (also in profile view), in face view elliptic or ovate; not dextrinoid, yellowish in KOH, wall thin and appearing smooth (or under oil immersion faintly rugulose).

Hymenium.—Basidia 4-spored. Pleurocystidia none. Cheilocystidia (30) 48–65 × 7–9 × 4–6 μm, narrowly fusoid-ventricose to cylindric-clavate; hyaline in KOH.

Lamellar and pilear tissues.—Lamellar trama of subparallel slender hyphae typical of the genus. Cuticle of pileus an ixotrichodermium collapsing to an ixolattice. Hypodermium cellular, ochraceous-brown in KOH. Pilear trama typical of the genus. Clamp connections present.

Habit, habitat, and distribution.—On soil under conifers, Independence Pass area, Pitkin County, Colorado, August 4, 1976, Smith 87043 (type, MICH).

Observations.—*H. nigromaculatum* is a second species which develops blackish spots on the pileus, but it is readily distinguished from *H. substrophosum* by the narrower stipe, hyphoid hypodermium and in having an ixocutis over the pileus. *H. substrophosum* appears to be closely related to *H. fastibile* sensu Horak, but again, as with *H. pseudostrophosum,* the presence of an ixotrichodermium over the pileus is distinctive.

16. Hebeloma vinaceoumbrinum sp. nov.

Illus. Figs. 11–12.

Pileus (1) 2–3 cm latus, late convexus vel umbonatus, glaber, viscidus, atrovinaceus dem obscure testaceus, ad marginem fibrillosus; odor et gustus raphaninus. Lamellae pallidae, tarde cinnamomeae, confertae, latae, adnatae. Stipes 4–7 cm longus, 3–5 mm crassus, albidus, brunnescens. Velum albidum, fibrillosum, evanescens. Sporae 7.5–10 × 5–6 μm,. non dextrinoideae, ellipsoideae vel ovoideae. Cheilocystidia 38–56 × 6–9 × 4–6 μm, obtusa. Cuticula pileorum ixotrichoderma est.

Specimen typicum in Herb. Univ. Mich. conservatum est; Smith 89950; legit sub *Alni,* prope Burnt Mt., Pitkin County, Colorado, 25 Aug 1979.

Pileus (1) 2–3 cm broad, convex becoming broadly convex, sometimes obtusely umbonate, disc glabrous and viscid, deep vinaceous-brown ("Natal Brown") becoming paler to "Army Brown" or redder; margin white fibrillose from remains of the veil, edge often fringed with fibrils. Context watery grayish, fading to white; odor and taste raphanoid; $FeSO_4$ on base of stipe olivaceous.

Lamellae broad, close, adnate, whitish at first as seen across the gills, faces at first avellaneous becoming dull cinnamon; edges not beaded and not spotted.

Stipe 4–7 cm long, 3–5 mm thick, equal, solid, white and fibrillose overall at first but veil white and not leaving an annulus; near base soon becoming brown and the change progressing upward (the $FeSO_4$ color change most pronounced on the discolored area).

Spores 7.5–10 × 5–6 μm, smooth in KOH, minutely punctate in Melzer's, not dextrinoid, in face view ovate to elliptic, in profile ovate to elliptic (*not* inequilateral to any degree).

Hymenium.—Basidia 4-spored, narrowly clavate, 7–9 μm broad, some with hyaline globules within as mounted in KOH. Pleurocystidia none. Cheilocystidia abundant, not agglutinated, cylindric down to a subventricose basal area, 38–56 × 6–9 × 4–6 μm, neck straight, not branched or forked.

Lamellar and pilear tissues.—Lamellar trama typical for the genus (not red in Melzer's). Cuticle of pileus an ixotrichodermium of long tubular elements, walls near hypoderm ± finely roughened, hyphae 2.5–3.5 μm diam, walls gelatinizing, clamps present. Hypodermium hyphoid, the hyphae 5–11 (15) μm wide, walls rusty brown in KOH and some incrustations present, no dextrinoid debris noted. Pileus trama of compactly interwoven hyaline to ochraceous, smooth-walled hyphae of various widths, lacking a red flush in Melzer's.

Habit, habitat, and distribution.—More or less clustered under alder, Snowmass Village, Pitkin County, Colorado, August 25, 1979 (type, MICH).

Observations.—The diagnostic characters are: (1) the heavy pallid veil, (2) the distinctly reddish brown pileus, (3) relatively straight-necked cheilocystidia, and the raphanoid odor and taste. The habitat may be misleading—there were conifers nearby. *H. collariatum* Bruchet is very close to our species but has larger spores and apparently less red in the pileus.

17. **Hebeloma pseudostrophosum** sp. nov.

Illus. Pl. 2; figs. 20–21.

Pileus 3–7 cm latus, obtusus deinde expanso-umbonatus, demum undulatus, glutinosus, ad marginem squamulosus, pallide fulvus demum argillaceus; odor et gustus raphaninus. Lamellae confertae, angustae demum latae, adnatae, pallidae demum fulvae. Stipes 5–8 cm longus, 8–18 mm crassus, deorsum attenuatus, fibrilloso-annulatus. Velum pallide argillaceum. Sporae 7–9 × 5–5.5 μm, ellipsoideae vel ovoideae, non dextrinoideae, subleves. Cheilocystidia 40–70 × 7–12 × 3–5 μm, fusoid-ventricosa. Cuticula pileorum ixotrichoderma est.

Specimen typicum in Herb. Univ. Mich. conservatum est, Smith 14443; legit prope Olympic National Park, Washington, 19 Jun 1939.

Pileus 3–7 cm broad, obtuse becoming broadly umbonate or at times the margin uplifted and the disc depressed, margin inrolled at first and wavy in age, slimy-viscid at first, the surface decorated with several rows of squamules from the broken heavy veil but these gradually becoming weathered away; color "Cinnamon" to "Cinnamon-Buff" or "Ochraceous-Tawny," the margin often paler and "Pinkish

Buff" to near avellaneous (grayer), the disc often dingy clay color in age. Context thick (± 6 mm in disc), tapered evenly to margin, pliant, watery-avellaneous, odor pungent-raphanoid, taste slightly raphanoid.

Lamellae narrow to moderately broad (4–6 mm), close, bluntly adnate, not seceding, pallid when young becoming "Vinaceous-Buff" and finally rusty brown; edges even or becoming eroded, not beaded.

Stipe 5–8 cm long, 8–18 mm thick, equal or narrowed downward, solid, lower two-thirds covered by "Cinnamon-Buff" patches or zones of veil remnants, pallid above, apex silky and concolorous with young gills, bister from base up in age. Veil heavy and composed of a cinnamon-buff outer layer and a white inner layer.

Spores 7–9 × 5–5.5 μm, ellipsoid to ovoid, pale clay color in KOH, not dextrinoid, surface obscurely marbled.

Hymenium.—Basidia 4-spored, 36–40 × 7–9 μm, clavate. Pleurocystidia none. Cheilocystidia 40–70 × 7–12 × 3–5 μm, the basal ventricose portion small at first and often disappearing in age, the neck elongate and apex obtuse, thin-walled, hyaline in KOH.

Lamellar and pilear tissues.—Lamellar trama typical for the genus. Cuticle of pileus an ixotrichodermium finally collapsing to an ixolattice, the hyphae 2–4 μm diam, clamp connections present. Hypodermium hyphoid, dingy clay colored as revived in KOH.

Habit, habitat, and distribution.—Gregarious under conifers, Olympic National Park, Washington, June 19, 1939 (type, MICH).

Observations.—This species cannot be identified with *H. strophosum* since Fries clearly stated (Monographia I, p. 325) that *H. strophosum* has a white veil. The diagnostic features for *H. pseudostrophosum* are: the two-layered veil, the veil being copious and leaving a ragged annular zone on the stipe, the small nondextrinoid spores, the cuticle of the pileus being an ixotrichodermium (and accompanying this the slimy pileus when fresh), and a raphanoid odor and taste. The cuticle of the pileus is best studied on young specimens. The presence of an ixotrichodermium over the pileus removes *H. pseudostrophosum* from *H. fastibile* sensu Ricken or as interpreted by Horak (1968). Horak illustrated the spores of the latter as distinctly ornamented. *H. versipelle* is also close to our species, but it was described as having a thin veil and lacking both a raphanoid odor and taste.

18. Hebeloma alpinicola sp. nov.

Illus. Figs. 25–27.

Pileus 2–4 cm latus, demum subplanus vel undulatus, ad marginem fibrillosus, glabrescens, glutinosus, cinnamomeus. Odor et gustus raphaninus Lamellae subdistantes, latae, demum fulvocinnamomeae. Stipes 3–6 cm longus, 4–8 (11) mm crassus, aequalis, albidus tarde brunnescens. Velum bicoloratum (ochraceum et albidum). Sporae 7–9 × 5–5.5 μm, ovoideae vel ellipsoideae, non dextrinoideae. Cheilocystidia 28–43 × 5–7 μm, fusoid-ventricosa vel clavata vel subcylindrica.

Specimen typicum in Herb. Univ. Mich. conservatum est, Smith 58632; legit sub *Pinus albicaulis*, prope Riggins, Idaho, 5 Jul 1958.

Pileus 2–4 cm broad, convex with an incurved margin, becoming broadly convex to plane, the margin spreading and ± wavy in age at times, ± evenly cinnamon in color, marginal area at first decorated with patches of veil fibrils but soon glabrous, viscid to thinly slimy in wet weather, opaque. Context pallid to brownish, odor and taste raphanoid.

Lamellae broad at maturity, moderately close to subdistant, adnate, seceding, at first pinkish buff but becoming pale rusty cinnamon, edges even, not beaded.

Stipe 3–6 cm long, 4–8 (11) mm thick, equal, white at first and decorated with veil remnants (the outer layer cinnamon-buff, the inner layer white to pallid); becoming clay color or darker from the base upward.

Spores 7–9 × 5–5.5 μm, ovoid to ellipsoid or in profile view some obscurely inequilateral, appearing smooth, nearly hyaline in KOH, weakly yellowish in Melzer's (not dextrinoid).

Hymenium.—Basidia 4-spored, 6–7.5 μm broad near apex. Pleurocystidia none. Cheilocystidia 28–43 × 5–7 × 7–9 μm, versiform: subcylindric, ± clavate, or obscurely fusoid-ventricose, hyaline as mounted in KOH, thin-walled, scattered.

Lamellar and pilear tissues.—Lamellar trama pale rusty cinnamon in sections mounted in KOH, otherwise typical for the genus (no red color in mounts in Melzer's). Cuticle of pileus an ixolattice (possibly an ixotrichoderm when very young), the elements 1.5–2.5 μm diam, refractive in KOH, clamped. Hypodermium cellular to hyphoid (intermediate as to type), pale fulvous in KOH, redder in Melzer's, incrustations scattered on the hyphal walls. Tramal hyphae ± ochraceous in KOH.

Habit, habitat, and distribution.—Gregarious under *Pinus albicaulis*, Heaven's Gate Ridge, Seven Devils Mts. near Riggins, Idaho, July 5, 1958 (type, MICH).

Observations.—The moderately copious bicolored veil and the radishlike odor and taste separate this species from *H. versipelle* where we first tried to place it. Its tree associate was clearly with the local alpine 5-needle pine. The cheilocystidia separate it from *H. bicoloratum* var. *coloradense,* a variety associated with spruce.

Stirps NIGROMACULATUM

Pileus and/or gills staining dark brown to blackish finally.

KEY TO SPECIES

1. Stipe readily splitting lengthwise; lamellae narrow; veil in young
basidiocarps cinnamon-buff .... *H. angustifolium* (c) (p. 164)
1. Not as above 2
   2. Veil grayish (and pileus at first gray-canescent from remains of the veil)
.... 19. *H. griseocanescens*
   2. Veil white to buff or tan 3
3. Veil whitish (pallid) (see *H. substrophosum*, p. 45) ...... 20. *H. nigromaculatum*
3. Veil (or at least outer layer) buff to pale tan 4
   4. Odor unpleasant (crush the context); taste mild; FeSO₄ staining
all parts green .... 21. *H. brunneomaculatum*
   4. Not as above 5
5. Hypodermium hyphoid; lamellae staining dark brown finally .. 22. *H. angelesiense*
5. Hypodermium cellular; lamellae not staining .... 23. *H. sublamellatum*

19. Hebeloma griseocanescens sp. nov.

Pileus 1–3.5 cm latus, obtusus demum convexus vel planus, griseocanescens, glabrescens, ad centrum atrobrunneus; odor et gustus mitis. Lamellae confertae, latae, adnatae, griseae demum obscure cinnamomeae, brunneomaculatae. Stipes 2–3.5 cm longus, 1.5–3 mm crassus, rufobrunneus, griseo-fibrillosus, tarde striatus. Velum cinereum, evanescens. Sporae 8–10 (11) × 5–6 μm, in "KOH" ochraceae, subleves, oblongae vel ellipsoideae, non dextrinoideae. Cheilocystidia versiformia: 18–27 × 8–15 μm, utriformia vel clavata; 36–40 × 8–11 × 4–6 μm, elongate fusoid-ventricosa, ± 50 × 6–8 μm, elongate clavata.

Specimen typicum in Herb. Univ. Mich. conservatum est, Smith 88697; legit prope Independence Pass, Pitkin County, Colorado, 23 Jul 1978, Evenson and Smith.

Pileus 1–3.5 cm broad, obtuse to convex, becoming plane, surface viscid but soon dry, grayish hoary at first from a thin coating of veil fibrils, blackish brown beneath the canescence, cinnamon-brown in age. Context concolorous with the moist surface, odor none, taste mild; FeSO₄ staining stipe base olive-black.

Lamellae broad, close, adnate, at first grayish, in age ± "Verona Brown," edges stained darker in age but not beaded (and stains not resulting from dried droplets).

Stipe 2–3.5 cm long, 1.5–3 mm thick, equal, tubular, cortex dull brown; surface reddish brown and decorated with a thin coating of grayish fibrils, soon dark brown overall to (finally) blackish and with only faint traces of the veil.

Spores 8–10 (11) × 5–6 μm, yellowish in KOH, nearly hyaline revived in Melzer's (not dextrinoid), faintly marbled, mostly oblong to ellipsoid, in profile view a very small number obscurely inequilateral.

Hymenium.—Basidia 4-spored, 7–8 μm broad. Pleurocystidia absent. Cheilocystidia versiform: utriform and 18–26 × 8–11 μm, clavate and 18–27 × 8–15 μm, elongate-fusoid-ventricose and 36–40 × 8–11 × 4–6 μm, and finally some elongate-clavate and ± 50 × 6–8 μm.

Lamellar and pilear tissues.—Lamellar trama typical for the genus. Cuticle of pileus an ixocutis fairly well developed, the hyphae 2.5–4 μm diam, walls refractive (gelatinized), clamps present. Hypodermium intermediate in type (a mixture of hyphal segments and cells) walls of the elements heavily encrusted and rusty ochraceous in KOH, redder brown in Melzer's. Tramal hyphae typical for the genus but with a tendency toward having encrusted walls especially near the hypodermium.

Habit, habitat, and distribution.—Gregarious on wet earth, Independence Pass area, Pitkin County, Colorado, July 23, 1978 (type, MICH).

20. Hebeloma nigromaculatum sp. nov.

Illus. Figs. 15–16.

Pileus (1.5) 2–4 cm latus, obtusus deinde convexus vel umbonatus, viscidus, ad marginem fibrillosus, subspadiceus deinde atromaculatus; odor et gustus raphaninus. Lamellae brunneolae deinde griseobrunneae, demum fulvomaculatae, confertae, latae, adnatae. Stipes 3–5 cm longus, 2.5–4 mm crassus, brunnescens. Velum pallidum, fibrillosum. Sporae 8–11 × 5–6 μm, non dextrinoideae. Cheilocystidia 50–80 × 8–11 × 4.5–7 μm, fusoide ventricosa, ad apicem obtusa.

Specimen typicum in Herb. Univ. Mich. conservatum est, Smith 19314; legit prope Rhododendron, Oregon, 1 Oct 1944.

Pileus (1.5) 2–4 cm broad, obtuse to convex with an inrolled margin, becoming plane or with a very slight umbo, surface viscid at first, fringed with pallid veil remnants or veil leaving a zone of patches near the margin, in age ± glabrescent, color evenly "Tawny-Olive" (dingy yellow-brown), in age developing blackish stains, margin often slightly paler than the disc. Context concolor with surface, bister in age, odor and taste raphanoid.

Lamellae moderately broad (15 mm) and close, slightly depressed around the stipe at maturity, pallid brownish when young, near "Wood

Brown" at maturity, often with dingy rusty brown stains in age, edges even.

Stipe 3–5 cm long, 2.5–4 mm thick at apex, equal or enlarged downward, pallid when young or with only a tinge of brown, becoming dingy brown from the base upward, appressed fibrillose from remains of a pallid veil, apex finely pruinose.

Spores 8–11 × 5–6 μm, elliptic to obscurely ovate in face view, in profile view subelliptic to obscurely bean-shaped, appearing smooth (under light microscope), nearly hyaline in KOH, not dextrinoid.

Hymenium.—Basidia 4-spored. Pleurocystidia none. Cheilocystidia 50–80 × 8–11 × 4.5–7 μm, apex obtuse, near base slightly ventricose, the neck greatly elongated, thin walled, smooth.

Lamellar and pilear tissues.—Lamellar trama hyaline or nearly so (typical for the genus). Cuticle of pileus an ixocutis, hyphae 2–4.5 μm diam, hyaline, walls refractive, clamps present. Hypodermium hyphoid.

Habit, habitat, and distribution.—Gregarious on moss in sandy soil, Rhododendron, Oregon, under pine and hemlock; October 1, 1944 (type, MICH).

Observations.—*H. nigellum* Bruchet has larger spores (11–13 × 6.5–7.5 μm) and a less sharply defined ixocutis. Both have very dark colored pilei, but in drying pilei of *H. nigromaculatum* are buff on the margin and ± pale "Verona Brown" over the disc—the blackening disappears.

21. Hebeloma brunneomaculatum sp. nov.

Illus. Fig. 27A.

Pileus 2–4 cm latus, convexus, deinde planus vel umbonatus, cinnamomeo-brunneus, demum ± aurantiobrunneus, subviscidus, demum brunneomaculatus, ad marginem fibrillosus. Velum pallide argillaceum. Odor distinctus sed non raphaninus vel fragrans. Lamellae latae, adnatae, confertae, brunneolae deinde obscure cinnamomeae. Stipes 4.5–5 cm longus, 4–7 mm crassus, brunneus, fibrillosus. Sporae 7–9 × 5–5.5 μm, in cumulis fulvae, ovoideae vel ellipsoideae, subleves, non dextrinoideae. Cheilocystidia 36–58 × 7–8 × 4–5.5 μm, ± fusoid-ventricosa.

Specimen typicum in Herb. Univ. Mich. conservatum est, Wells-Kempton 7; legit prope Matanuska River Campground, Alaska, 20 Aug 1966.

Pileus 2–4 cm broad, convex with the margin connivent, expanding to plane or retaining a low umbo, hygrophanous, when moist cinnamon-brown, distinctly paler when faded and a medium orange-brown, subviscid when moist, usually spotted with dark brown in age, near edge at first thinly coated with pale tan veil fibrils. Context pallid, 2–4 mm thick in disc, odor rather unpleasantly acrid (like crushed daffodil stems at first or of fern leaves), taste not distinctive; with $FeSO_4$ greenish on all parts.

Lamellae moderately broad (3–5 mm), close, adnate to adnexed, tan becoming dull brown, edges slightly crenate and concolorous with faces, not beaded.

Stipe 4.5–5 cm long, 4–7 mm thick, equal, ivory to tan becoming dark brown from the base upward, dry, apex pruinose, longitudinally appressed fibrillose, hollow; veil pale tan, fibrillose, scanty and soon vanishing.

Spore deposit dull rusty brown fresh, as air-dried a dull clay color. Spores 7–9 × 5–5.5 μm, smooth in KOH, ovoid to ellipsoid, in KOH nearly hyaline singly, about the same in Melzer's (not dextrinoid).

Hymenium.—Basidia 4-spored, 8–9 μm broad, projecting when sporulating. Pleurocystidia none. Cheilocystidia 36–58 × 7–8 × 4–5.5 μm, subcylindric to narrowly fusoid-ventricose, apex obtuse, hyaline, thin-walled, slightly refractive (viscid ?).

Lamellar and pilear tissues.—Lamellar trama typical for the genus but flushed reddish when first mounted in Melzer's. Cuticle of pileus a thick ixocutis, the hyphae 2–4 μm, walls refractive, clamps present but not abundant. Hypodermium typically hyphoid and ± indefinite in extent, dull clay color in KOH, slightly redder in Melzer's from a diffused pigment. Tramal hyphae not distinctive, clamps rare.

Habit, habitat, and distribution.—Gregarious in moss under spruce, Matanuska River Campground, Alaska, August 20, 1966, Wells-Kempton (type, MICH).

Observations.—This species is close to *H. pseudostrophosum* but it is distinct because of the peculiar odor and the cuticle of the pileus being an ixocutis, not an ixotrichodermium. Also, the veil is much thinner.

22. Hebeloma angelesiense sp. nov.

Illus. Pl. 1; figs. 19, 22.

Pileus 2–6 cm latus, demum plano-umbonatus, viscidus, cinnamomeus deinde argillaceus, ad marginem avellaneus, fibrillosus deinde glabrescens. Odor et gustus raphaninus. Lamellae latae, confertae, adnatae avellaneae demum subcinnamomeae et brunneomaculatae. Stipes 5–8 cm longus, 8–12 mm crassus, deorsum attenuatus, brunnescens, lacerate fibrillosus. Velum bicoloratum (argillaceum et albidum).

Sporae 7–9.5 × 5–6 μm, ovoideae vel obscure inequilaterales, subdextrinoideae. Cheilocystidia 40–60 × 4–5 μm, ventricosa vel filamentosa.

Specimen typicum in Herb. Univ. Mich. conservatum est, Smith 17096; legit prope Port Angeles, Washington, 21 Sep 1941.

Pileus 2–6 cm broad, obtuse and with an incurved margin at first, expanding to nearly plane with a wavy margin and obtuse umbo, the margin rarely uplifted, marginal area with scattered patches of appressed fibrils from the remains of the veil, surface viscid and fibrillose-streaked beneath the slime, glabrescent in age, "Cinnamon" on disc and paler over the margin, becoming dingy clay color on disc and near "Avellaneous" (grayer) along the margin, disc darker when water-soaked. Context thick (about 6 mm), tapered evenly to cap margin, pliant, watery avellaneous, odor pungent, taste slightly raphanoid.

Lamellae only moderately broad but becoming slightly ventricose (5–6 mm), close, depressed-adnate or with a slight tooth, 1–3 tiers of lamellulae, pale avellaneous at first becoming "Sayal Brown" in age, staining darker on the edges where injured, edges becoming eroded, not beaded.

Stipe 5–8 cm long, 8–12 mm thick, equal or narrowed downward, hollow, pallid in cortex, surface coarsely appressed or ragged-fibrillose from the bicolored veil featuring an argillaceous outer and a white inner layer; surface pallid near apex, cinnamon-buff over the remainder and becoming bister from the base upward; annular zone evanescent.

Spores 7–9.5 × 5–6 μm, ovate to elliptic in face view, elliptic to slightly bean-shaped or obscurely inequilateral in profile, smooth under a high-dry objective, very minutely marbled under oil-immersion lens, nearly hyaline in KOH, reddish tawny in Melzer's (but not truly dextrinoid).

Hymenium.—Basidia 4-spored, 20–24 × 7–8 μm. Pleurocystidia none. Cheilocystidia narrowly fusoid-ventricose, 40–60 × 6–9 × 4–5 μm, some filamentous in age, thin-walled, hyaline in KOH.

Lamellar and pilear tissues.—Lamellar trama typical for the genus. Cuticle of pileus an ixocutis of hyphae 2–4 μm diam, walls refractive and clamp connections present. Hypodermium hyphoid, when fresh nearly hyaline in KOH. Tramal hyphae typical for the genus.

Habit, habitat, and distribution.—Gregarious to subcaespitose near conifers, Mt. Angeles, Port Angeles, Washington, September 21, 1941 (type, MICH).

Observations.—The bicolored veil with remains copious enough to cause the stipe to be ragged-fibrillose below the veil line, the relatively thick stipe for a species in this group, and the generally dingy cinnamon color of the pileus are distinctive.

23. Hebeloma sublamellatum sp. nov.

Pileus 2.4–4 cm latus, late convexus, glaber, ad centrum obscure rubrobrunneus, ad marginem griseobrunneus, saepe virgatus, demum brunneomaculatus. Velum sparsim, fibrillosum. Odor et gustus raphaninus. Lamellae pallidae demum obscure rubrobrunneae, confertae, latae, adnatae. Stipes 3–6 cm longus, 4–9 mm crassus, ad basin brunneus. Sporae 8–10 (11) × 5–6 μm, phaseoliformes vel ovoideae vel oblongae, subleves. Cheilocystidia 52–68 × 7–9 × 4–6 μm, anguste fusoid-ventricosa.

Specimen typicum in Herb. Univ. Mich. conservatum est, Smith 87044; legit prope Independence Pass, Pitkin County, Colorado, 4 Aug 1976, Mrs. Robert Scates.

Pileus 2.4–4 cm broad, broadly convex, the margin inrolled, surface "Verona Brown" on disc and grayish brown over the marginal area which is often virgate, tending to stain bister or darker over disc in age, margin decorated with veil remnants for only a short time. Veil ± pinkish buff. Context watery brown, fading to white; odor and taste raphanoid, $FeSO_4$ greenish on the stipe, KOH no reaction.

Lamellae moderately broad, close, adnexed, edges even, surface pallid then pinkish brown ("Verona Brown"), not staining.

Stipe 3–6 cm long, 4–9 mm thick, equal, solid, pallid within but brownish in the base; surface with patches of the buff-colored veil variously disposed, cortex slowly darkening from the base upward.

Spores 8–10 (11) × 5–6 μm, in profile obscurely bean-shaped to oblong, in face view elliptic to ovate, appearing smooth under a high-dry lens, in KOH brownish, not dextrinoid.

Hymenium.—Basidia 4-spored, 10–12 μm broad near apex. Pleurocystidia none. Cheilocystidia 52–68 × 7–9 × 4–6 μm, ventricose near base and with an elongated neck.

Lamellar and pilear tissues.—Lamellar trama typical for the genus. Cuticle of pileus an ixocutis, the hyphae 2–3 μm diam and yellowish in KOH, clamp connections present. Hypodermium a thick layer of rusty brown cells as revived in KOH.

Habit, habitat, and distribution.—Scattered under conifers, Independence Pass, Pitkin County, Colorado, August 4, 1976; Mrs. Robert Scates (type, MICH, Smith 87044).

Observations.—This species is distinct from *H. luteobrunneum* in having an ixocutis over the pileus, in having a raphanoid odor and taste, and in having a pileus which becomes ± spotted darker in age. It may eventually be found desirable to regard it as a variety of *H. mesophaeum.*

Stirps PASCUENSE

Stipe not darkening at the base.

KEY TO SPECIES

1. Spores 7–9 × 4.5–5 μm; odor mild; cheilocystidia 26–37 (44) × 6–9
   × 3–5 μm; under spruce; pileus hoary at first and only slightly viscid
   when mature .... *H. urbanicola* (r) (p. 181)
1. Not as above 2
   2. Odor and taste strongly raphanoid; stipe readily splitting lengthwise
   .... *H. proximum* (k) (p. 173)
   2. Not as above 3
3. Pilear cuticle an ixocutis; stipe 2–5 mm thick; stipe white overall . 24. *H. kauffmanii*
3. Not as above 4
   4. Hypodermium colorless in KOH; odor and taste more or less raphanoid
   .... *H. pascuense* (j) (p. 172)
   4. Not as above 5
5. Pileus rimose in age; spores up to 7 μm wide .... *H. gregarium* (e) (p. 167)
5. Not as above 6
   6. Stipe 3–5 mm thick, cinnamon-buff; walls of hypodermial
   hyphae with conspicuous incrustations (as revived in KOH)
   .... 25. *H. perigoense*
   6. Stipe 8–12 mm thick; hypodermial hyphae brown but incrustations
   not prominent (if present) under a 54 × oil-immersion lens
   .... 26. *H. subargillaceum*

24. Hebeloma kauffmanii sp. nov.

Pileus 1.5–3 cm latus, convexus vel obtusus demum ± planus vel obtuse umbonatus, ad centrum subfulvus, ad marginem ± alutaceus. Contextus albus fragilis, odor nullus, sapor subnauseosus. Lamellae demum emarginatae, confertae latae, cinnamomeae. Stipes 2–4 cm longus, 3–4 mm crassus, solidus demum cavus, pallidus sericeus. Velum sparsum, fibrillosum, album. Sporae 8–10 (11) × 5–6.5 μm, leves, subinequilaterales, non dextrinoideae. Cheilocystidia filamentosa, cylindrica vel fusoide ventricosa, 38–63 × 3–7 × 2–3.5 μm subacuta, obtusa vel (rare) subcapitata.

Specimen typicum in Herb. Univ. Mich. conservatum est, Kauffman 9-7-23; legit prope Centennial, Wyoming.

Pileus 1.5–3 cm broad, convex to obtuse, then ± plane or with an obtuse umbo, alutaceous or (on umbo) slightly tawny, subviscid, pellicle present; surface even, margin at first slightly white-cortinate. Context white when faded, thin, fragile, odor none, taste slight to subnauseous (Kauff. notes).

Lamellae adnexed, deeply emarginate, moderately broad, becoming ventricose, close but not crowded; edges undulating and at first white-fimbriate, faces cinnamon-buff to darker.

Stipe 2–4 cm long, 3–4 mm thick, equal, stuffed becoming hollow, slightly silky from a thin cortina, white overall and merely dingy in age, silky-fibrillose.

Spores 8–10 (11) × 5–6.5 μm, nearly hyaline in KOH, pale alutaceous on standing, about the same in Melzer's, smooth, ovate to elliptic in face view, in profile the ventral line ± straight, dorsal line convex, apex ± rounded to obtuse, not dextrinoid.

Hymenium.—Basidia 4-spored. Pleurocystidia none. Cheilocystidia filamentous to cylindric, subaciculate to narrowly fusoid-ventricose, 38–63 × 3–7 × 2–3.5 (5) μm, apices subacute, obtuse or rarely subcapitate, thin-walled, hyaline in KOH.

Lamellar and pilear tissues.—Lamellar trama typical of the genus (not red in Melzer's). Cuticle of pileus a thin ixocutis, the hyphae 2–3.5 μm diam. Hypodermium intermediate as to type, walls of elements tawny in KOH, not or only inconspicuously encrusted (layer appearing "cellular" in tangential sections). Tramal hyphae typical of the genus. Clamps present.

Habit, habitat, and distribution.—Gregarious to subcaespitose on black soil along a stream in a pine forest, Medicine Bow Mts., Wyoming (above Centennial), September 7, 1923, C. H. Kauffman.

Observations.—*H. kauffmanii* differs from *H. gregarium* Peck in not having either a raphanoid odor or taste, in the pileus not becoming radiately rimose, and in having an intermediate type of hypodermium which in KOH is fulvous brown and in which (in tangential sections) the elements are greatly inflated.

From *H. pascuense*, *H. kauffmanii* is readily distinguished by its "cellular" hypodermium (in tangential sections), in having the walls of the hypodermial elements highly colored (fulvous) in KOH, and by the lack of a raphanoid odor and taste.

25. Hebeloma perigoense sp. nov.

Pileus 1.5–4.5 cm latus, convexus deinde late convexus, subviscidus, ad marginem fibrillose squamulosus, glabrescens, luteobrunneus, ad centrum obscurior, odor et gustus ± raphaninus. Lamellae adnatae, confertae, latae, obscure cinnamomeae. Stipes 3–6 cm longus, 3–5 mm crassus, ad apicem fibrillose zonatus vel annulatus, deorsum non brunnescens. Velum pallide argillaceum. Sporae 8.5–10 × 5–5.5 μm, oblongae vel ellipsoideae vel phaseoliformes, non dextrinoideae. Cheilocystidia 30–55 × 7–9 μm, clavata vel filamentosa.

Specimen typicum in Herb. Denver Bot. Garden conservatum est, DBG 4877; legit prope Perigo, Gilpin County, Colorado, 13 Aug 1974, Chapman, Mitchel, and Smith.

Pileus 1.5–4.5 cm broad, convex to plane, surface somewhat viscid, margin at first decorated with remnants of the dingy buff-colored veil, disc glabrous, colors in the yellow-brown range, darker on disc and paler on margin. Odor and taste ± raphanoid.

Lamellae broad, close to subdistant, adnate, pinkish buff when young, near "Sayal Brown" at maturity, edges not beaded and not spotted.

Stipe 3–6 cm long, 3–5 mm thick, near the apex with a definite zone of fibrils or a fibrillose annulus, color of stipe ± "Cinnamon-Buff" overall from early youth on (not darkening at the base), only moderately fragile.

Spores 8.5–10 × 5–5.5 μm, oblong to ellipsoid, ± smooth in KOH, some containing a large hyaline oil droplet, in profile obscurely bean-shaped to elliptic, not dextrinoid (yellowish in Melzer's).

Hymenium.—Basidia 2- and 4-spored, 20–24 × 5–8 μm, clavate, many with small oil droplets in enlarged portion. Pleurocystidia absent. Cheilocystidia 30–55 × 7–9 μm, filamentous to clavate, rarely with a slight basal enlargement (5–10 μm diam), many ± cylindric-capitate.

Lamellar and pilear tissues.—Lamellar trama typical for the genus. Cuticle of pileus an ixocutis, the hyphae 3–6 μm diam, interwoven, hyaline to pale yellow in KOH, clamp connections present. Hypodermium cellular, the layer dark brown in KOH, pigment incrustations typically numerous and conspicuous on the hyphal walls. Hyphae of the pileus trama typical for the genus.

Habit, habitat, and distribution.—Gregarious under conifers, Perigo, Gilpin County, Colorado, August 13, 1974; legit Chapman, Mitchel, and Smith (type, DBG). The species is also known from Payette Lakes, Idaho, Valley County, June 30, 1954, Smith 44396.

Observations.—The incrustations on the hypodermial elements in Smith 44396 were much less pronounced than in the type, but we hesitate to emphasize this difference taxonomically.

26. Hebeloma subargillaceum sp. nov.

Pileus 7–25 mm latus, convexus demum late convexus vel ad centrum leviter depressus, ochraceus, viscidus, ad marginem albofibrillosus, glabrescens; odor et gustus raphaninus. Lamellae confertae, latae, adnatae, avellaneae, serrulatae. Stipes 3–7 cm longus, 8–12 mm crassus, ochraceus, sursum pruinosus, deorsum fibrillosus vel squamulosus. Velum copiosum, albidum, fibrillosum, evanescens. Sporae 8.5–10 × 5–5.5 μm, leves, ellipsoideae vel ovoideae, non dextrinoideae. Cheilocystidia 60–70 × 7–12 × 4–6 μm, fusoid-ventricosa, subacuta vel obtusa.

Specimen typicum in Herb. Denver Bot. Garden conservatum est, DBG 7947; legit prope Platero, Conejos County, Colorado, 24 Aug 1978, Vera Evenson.

Pileus 7–25 mm broad, becoming broadly convex to slightly depressed, ochraceous overall or ± pinkish cinnamon over marginal area, at first the margin fibrillose from remains of the veil, viscid but soon dry. Context white, thin, not staining, odor and taste raphanoid; $FeSO_4$ no reaction.

Lamellae broad, close, adnate, horizontal, serrulate, avellaneous when young, not beaded.

Stipe 3–7 cm long, 8–12 mm thick at apex, ochraceous, not darkening, longitudinally twisted-striate, pruinose at apex, not darkening at base, with a superior fibrillose evanescent annulus (which may be rusty brown from spores caught in it).

Spores 8.5–10 × 5–5.5 μm, smooth, ellipsoid to ovoid, in profile some more or less inequilateral, not dextrinoid, ochraceous in KOH-mounts.

Hymenium.—Basidia 2- and 4-spored mixed, 18–20 × 7–9 μm. Pleurocystidia none. Cheilocystidia 60–70 × 7–12 × 4–6 μm fusoid-ventricose, apex subacute, hyaline.

Lamellar and pilear tissues.—Lamellar trama typical of the genus. Cuticle of pileus an ixolattice, the hyphae 2–3 μm diam, with clamp connections. Hypodermium hyphoid, rusty brown in KOH, the hyphae 3–6 μm diam.

Habit, habitat, and distribution.—In grass under spruce, Green Lake area south of Platero, Conejos County, Colorado, August 24, 1978 (type, DBG 7947).

Observations.—The important characters for this species include the following: (1) the thick stipe (8–12 mm); (2) the stipe not darkening; (3) the odor and taste are raphanoid; (4) a pinkish cinnamon pileus; (5) a simple ixocutis over the pileus; (6) large fusoid-ventricose cheilocystidia with at least some of them with the apex acute; and (7) the rusty brown color of the hypodermium as revived in KOH. It is very close to *H. fastibile* but differs in slightly smaller ellipsoid spores and the evenly pale-colored pileus when young and fresh. The spores appear smooth under a high-dry, 4-mm objective which is quite different from the distinct ornamentation of the spores as illustrated by Horak for *H. fastibile*.

Stirps DISSILIENS

Stipe splitting lengthwise.

27. Hebeloma dissiliens sp. nov.

Pileus 1.5–3.5 (4) cm latus, demum plano-umbonatus, subviscidus, ad marginem fibrillosus obscure cinnamomeus ("Sayal Brown"); contextus insipidus, odor nullus. Lamellae brunneolae demum obscure cinnamomeae, confertae, angustae, adnatae. Stipes 3–6 cm longus, 2–3 mm crassus, dissiliens, deorsum demum obscure fulvus; velum albidum, sparsim, fibrillosum. Sporae 9–11.5 × 4.5–5.5 μm, anguste ellipticae vel subphasioliformes, non dextrinoideae, subleves. Cheilocystidia anguste fusoide ventricosa vel subcylindrica, 32–48 × 4–7 × 3–6 μm.

Specimen typicum in Herb. Univ. Mich. conservatum est, Smith 89723; legit prope Elk Camp, Burnt Mt., Pitkin County, Colorado, 12 Aug 1979.

Pileus 1.5–3.5 (4) cm broad, obtuse, becoming nearly plane or plano-umbonate, margin in some uplifted, for a time with patches of veil remnants along the margin causing it to be pallid, disc (and finally overall) cinnamon-brown to dingy cinnamon ("Sayal Brown"). Context thin, soft, brownish fading to pallid, odor and taste mild. FeSO$_4$ staining stipe base olive-black.

Lamellae pallid brown becoming dull cinnamon, crowded, adnate, narrow, not beaded or spotted.

Stipe 3–6 cm long, 2–3 mm thick, narrow, readily splitting lengthwise, soon dull rusty brown from the base up, pallid over surface from a thin coating of whitish fibrils but no annular zone evident.

Spores 9–11.5 × 4.5–5.5 μm, nearly hyaline singly in KOH, in Melzer's nearly hyaline (not dextrinoid) and only minutely punctate, in face view narrowly ovate to ± elliptic, in profile obscurely bean-shaped to elliptic or obscurely inequilateral, as revived in KOH with a large globule in the interior.

Hymenium.—Basidia 4-spored. Pleurocystidia none. Cheilocystidia 32–48 × 4–7 × 3–6 μm, bent like a hockey stick at the base (or resembling a bent knee), scarcely ventricose near the base, ± cylindric.

Lamellar and pilear tissues.—Lamellar trama flushed rose color in Melzer's but soon fading and hymenium ± remaining reddish orange. Cuticle of pileus an ixocutis, the hyphae 3–5 μm diam and clamped at the septa, hyaline. Hypodermium cellular, rusty brown in KOH, in Melzer's with a reddish tone, hyphal walls incrusted. Tramal hyphae typical of the genus.

Habit, habitat, and distribution.—Gregarious under a mixture of spruce and fir with scattered lodgepole pine, Elk Camp, Burnt Mt., Pitkin County, Colorado, August 12, 1979 (type, MICH).

Observations.—The spore shape is intermediate between the bean-shaped and the inequilateral types as seen in profile view. The species, clearly, is close to *H. mesophaeum*. The distinguishing features are the fragile, readily splitting stipe, narrow gills, mild odor and taste, and the dull cinnamon pileus at maturity.

Stirps MESOPHAEUM

Stipe darkening, not splitting.

KEY TO SPECIES

28. Hebeloma sanjuanense sp. nov.

Pileus 2–4 cm latus, convexus, demum planus vel expando-umbonatus, ad centrum glaber, ad marginem argillaceo-fibrillosus, ad centrum ± rubrobrunneus, odor et gustus raphaninus. Lamellae adnatae, confertae, latae, obscure cinnamomae. Stipes 4–6 cm longus, 3–5 mm crassus, brunnescens. Velum sparsim, pallide argillaceum. Sporae 8.5–10 × 5–5.5 μm, subleves, phaseoliformes, ovoideae, vel ellipsoideae, non dextrinoideae. Cheilocystidia 30–45 × 5–7 μm, filamentosa, flexuosa, inconspicua.

Specimen typicum in Herb. Univ. Mich. conservatum est, Smith 51736; legit prope Lizard Head Butte, San Juan County, Colorado, 3 Aug 1956.

Pileus 2–4 cm broad, obtuse to convex becoming ± plane or with a slight umbo, glabrous over disc, with cinnamon-buff veil remnants along the margin at first, color over disc a dull vinaceous-brown ("Verona Brown"); context thin, odor and taste raphanoid.

Lamellae adnate, close, broad, at first with a vinaceous tint (in artificial light), soon ± "Sayal Brown," edges even, not beaded.

Stipe 4–6 cm long, 3–5 mm thick, equal, fragile, soon darkening from base upward; veil remnants soon vanishing.

Spores 8.5–10 × 5–5.5 μm, pale yellow in KOH and appearing smooth, ovate to elliptic in face view, ± bean-shaped in profile to subelliptic, rarely obscurely inequilateral, not dextrinoid.

Hymenium.—Basidia 4-spored, 24–29 × 6–8 μm. Pleurocystidia none. Cheilocystidia 30–45 × 5–7 μm, filamentous, flexuous, hyaline, inconspicuous.

Lamellar and pilear tissues.—Lamellar trama typical for the genus. Cuticle of pileus an ixolattice, elements 2–4 μm diam, often branched, some encrusted with hyaline to brown pigment. Hypodermium cellular, cell walls heavily encrusted and pigmented. Clamp connections present.

Habit, habitat, and distribution.—Scattered under conifers, Lizard Head Butte, near Teluride, San Juan County, Colorado, August 3, 1956 (type, MICH).

Observations.—This species differs from *H. mesophaeum* in the shape and size of the cheilocystidia, in the thicker more gelatinous cuticle of the pileus and in the color of the veil. *H. pumilum* Lange has small cheilocystidia but has a bitter rather than raphanoid taste and was collected under *Fagus*. Also, Lange's illustration of *H. pumilum* (1938–40, pl. 119B) indicates a white veil and spores inequilateral in profile view.

29. Hebeloma strophosum (Fr.) Saccardo var. occidentale var. nov.

Pileus 1.5–3 cm latus, late umbonatus, subviscidus, valde fibrillosus, tarde subglabrescens, pallide argillaceus; odor et gustus raphaninus. Velum pallidum demum pallide argillaceum. Lamellae confertae, latae, adnatae, pallidae demum cinnamomeae. Stipes 3–5 cm longus, 3–6.5 mm crassus, deorsum brunnescens, multizonatus vel subannulatus. Sporae 8.5–10.5 × 5–5.5 μm, Cheilocystidia 28–53 (65) × 4–10 μm.

Specimen typicum in Herb. Univ. Mich. conservatum est, Smith 44190; legit sub *Piceae,* prope McCall, Idaho, 25 Jun 1954.

Pileus 1.5–3 cm broad, obtuse, expanding to broadly umbonate, margin incurved and heavily fibrillose from the pallid veil, surface opaque, scarcely viscid when collected, color pale pinkish tan on disc and pinkish buff on the margin; context pallid, odor and taste raphanoid.

Lamellae close, broad, adnate, pallid when young, about "Sayal Brown" to dull tawny at maturity, neither beaded nor spotted.

Stipe 3–5 cm long, 3–6.5 mm thick, equal, not readily splitting, pallid above, yellowish and becoming darker (to dull brown) below. Veil pallid, very pale buff as dried, copious, often leaving zones or patches on the surface stipe ± annulate.

Spores 8.5–10.5 × 5–5.5 μm, appearing smooth, wall thin, shape in face view oblong to ± ovate, in profile ± oblong to elliptic, rarely obscurely inequilateral, not dextrinoid, in KOH very pale brownish.

Hymenium.—Basidia mostly 4-spored, 8–9 (11) μm wide near apex, hymenium reddish to ochraceous in Melzer's. Pleurocystidia none. Cheilocystidia 28–53 (65) × 4–10 μm, filamentous, elongate-clavate, or ventricose near base and with a long often flexuous neck (this type most numerous).

Lamellar and pilear tissues.—Lamellar trama typical for the genus but red as revived in Melzer's. Cuticle of pileus a poorly formed ixocutis, or "apparently" a cutis (the latter from veil hyphae adhering to the surface), the hyphae 2–4 μm wide, tubular and ± hyaline in KOH, clamp connections present. Hypodermium intermediate in type, the hyphae not or only slightly encrusted, the layer merely brown as revived in KOH. Tramal hyphae red in Melzer's and both dextrinoid and amyloid debris present in significant amounts.

Habit, habitat, and distribution.—Gregarious on soil along a forest road, Engleman spruce nearby, McCall, Idaho, June 25, 1954, Smith 44190.

Observations.—A collection by Kauffman (August 27, 1923, Medicine Bow Mts., Wyoming) apparently belongs here. In the type collection the cheilocystidia have a tendency to be slightly enlarged at the apex, a feature we now consider important in the genus. European authors have pointed out that *H. mesophaeum* and *H. strophosum* intergrade.

30. Hebeloma bicoloratum var. coloradense var. nov.

Illus. Pl. 10; figs. 23–24.

Pileus 2–4 cm latus convexus demum plano-expansus, viscidus, pallidofibrillosus, glabrescens, griseus vel brunneogriseus ("Drab" vel "Wood Brown") tarde demum brunneo-cinnamomeus, ad marginem ± appendiculatus. Contextus pallidus, odor et gustus subpungens (non raphaninus). Lamellae latae, confertae, adnatae, pallidae demum obscure cinnamomeae, non maculatae. Stipes 4–6 cm longus, 6–10 crassus, pallidus, fibrillosus, deorsum brunnescens. Sporae 8.5–11 × 5.5–6.5 (7) μm, ovoideae vel ellipsoideae vel in laterales subphasaeoliformes, non dextrinoideae. Cheilocystidia 34–75 × 4–7 μm, subclavata, filamentosa vel cylindrica vel fusoide-ventricosa.

Specimen typicum in Herb. Univ. Mich. conservatum est, Smith 89594; legit prope Elk Camp, Pitkin County, Colorado, 29 Jul 1979.

Pileus 2–4 cm broad, convex becoming plane or nearly so, ± viscid at first, coated with pallid veil fibrils, dark grayish brown (± "Drab" to "Wood Brown,") glabrescent and slowly fading to dull cinnamon ("Sayal Brown") or dingy tawny, patches of the veil persistent along the margin. Context pallid when faded, watery gray when moist, odor and taste ± pungent (not truly raphanoid); FeSO$_4$ quickly staining base of stipe olive to olive-fuscous.

Lamellae broad, close but distinct, adnate-seceding, pallid at first then becoming dull brown and finally cinnamon, thin, edges even and neither beaded nor stained dark brown.

Stipe 4–6 cm long, 6–10 mm thick, equal, pallid, with copious veil remnants and these showing an outer pinkish buff layer and a pallid inner one; surface soon becoming bister from the base upward.

Spores 8.5–11 × 5.5–6.5 (7) μm, ovate to elliptic in face view, elliptic to ± bean-shaped in profile, smooth as viewed in KOH, in Melzer's minutely marbled; clay color in KOH, thin-walled, not dextrinoid.

Hymenium.—Basidia 4-spored, 7–8 μm wide near apex, clavate. Pleurocystidia none. Cheilocystidia 34–75 × 4–7 μm, abundant, mostly elongate-subclavate to filamentous or cylindric, some ventricose at the base, hyaline and thin-walled.

Lamellar and pilear tissues.—Lamellar trama typical of the genus but in Melzer's ochraceous-orange and with copious dextrinoid debris precipitating in the mount. Cuticle of pileus an ixocutis, the hyphae 2–4 μm diam, tubular, walls clearly defined and the hyphae imbedded in the slime, walls finely roughened to encrusted near the hypodermium. Hypodermium hyphoid, dark rusty brown in KOH and in Melzer's, copious dextrinoid debris evident in the layer, incrustations in KOH rusty brown and conspicuous. Hyphae of pileal trama typical for the genus but red in Melzer's and dextrinoid debris copious.

Habit, habitat, and distribution.—Gregarious under spruce, Elk Camp, Pitkin County, Colorado, July 29, 1979, Smith 89594 (type, MICH).

Observations.—The veil remnants on the margin of the pileus do not always show the colored layer as distinctly as do the remains on the stipe in specimens in which the veil is about to break. The color of the pileus distinguishes it from *H. versipelle* sensu Romagnesi (1965) and from *H. sanjuanense.*

31. Hebeloma glabrescens sp. nov.

Pileus 1–3 cm latus, obtuse conicus, deinde late expansus vel umbonatus "Prout's Brown" deinde fulvus, ad marginem fibrillosus, glabrescens; odor pungens, gustus submitis. Lamellae pallidae demum cinnamomeae, confertae, latae, adnatae. Stipes 3–6 cm longus, 3–5 mm crassus, brunnescens, lacerato-fibrillosus. Velum griseoargillaceum. Sporae 8–10 (10.5) × 4.3–5 μm, ellipsoideae vel ovoideae, non dextrinoideae, subleves, pallide lutea cum "KOH." Cheilocystidia 40–80 × 6–12 μm, fusoid-ventricosa, elongata.

Specimen typicum in Herb. Univ. Mich. conservatum est, Smith 89053; legit prope Elk Wallow, Pitkin County, Colorado, 19 Aug 1978.

Pileus 1–3 cm broad, obtuse to convex becoming broadly expanded, viscid, margin decorated for a time with remains of the pale buff veil; color ± blackish brown ("Prout's Brown") when moist, when faded tawny, opaque at all times; odor pungent, taste slight, $FeSO_4$ on stipe base olive-gray.

Lamellae broad close, adnate, pallid becoming dull cinnamon, not beaded.

Stipe 3–6 cm long, 3–5 mm at apex, equal, darkening from the base up, shaggy-fibrillose from veil, veil grayish buff, an annular zone sometimes present.

Spores 8–10 (10.5) × 4.3–5 μm, elliptic to ovate in face view, in profile subelliptic, thin-walled, not dextrinoid, appearing smooth or under oil ± rugulose, merely yellowish in KOH.

Hymenium.—Basidia 4-spored, 17–25 × 6–12 μm. Pleurocystidia none. Cheilocystidia 40–80 × 6–12 μm, elongate fusoid-ventricose, apex obtuse to slightly enlarged, slightly ochraceous in the ventricose part.

Lamellar and pilear tissues.—Lamellar trama typical for the genus. Cuticle of the pileus a well-developed ixolattice of slender (2–3 μm diam) tubular hyphae. Hypodermium cellular, cell walls heavily incrusted, dark yellowish brown in KOH. Pileus trama compactly interwoven, the hyphae 5–9 μm diam, pallid ochraceous to hyaline in KOH. Clamps present.

Habit, habitat, and distribution.—Scattered under lodgepole pine, spruce, and fir, Fryingpan River above Elk Wallow, Pitkin County, Colorado, August 19, 1978 (type, MICH).

Observations.—The relatively narrow spores, grayish-buff copious veil, and the walls of the hypodermial cells heavily encrusted as mounted in KOH, are distinctive. *H. glabrescens* when fresh and moist has the colors of *H. mesophaeum* var. *holophaeum* Fries, but to properly settle the question of their identity a neotype must be selected for the Friesian variety. We cannot do this on the basis of North American specimens.

32. Hebeloma subviolaceum sp. nov.

Illus. Figs. 13–14.

Pileus 2.5–4 cm latus, late convexus demum ± planus, subviscidus, canescens, griseobrunneus demum rubrobrunneus vel atrobrunneus, squamulosus, glabrescens. Velum albidum. Odor and gustus mitis. Lamellae latae, confertae, adnatae, "Vinaceous-Buff" demum "Verona Brown," Stipes 3–5 cm longus, 4–6 mm crassus, fragilis, brunnescens, deorsum fibrillosus, subannulatus. Sporae 7–10 × 5–6 μm, subleves, ellipsoideae vel ovoideae, apice rotundatae, rare obscure inequilaterales. Cheilocystidia 26–35 × 10–15 μm, saccata vel ellipsoidea vel fusoid-ventricosa et 38–57 × 7–10 × 4–5 μm, tarde agglutinata.

Specimen typicum in Herb. Univ. Mich. conservatum est, Smith 88843; legit prope Elk Camp, Burnt Mt., Pitkin County, Colorado, 2 Aug 1978.

Pileus 2.5–4 cm broad, broadly obtuse becoming broadly convex or the margin uplifted; subviscid; color ± "Benzo Brown" beneath the veil fibrils, in age a dark "Verona Brown" to "Dresden Brown" or blackish brown ("Mummy Brown"). Veil whitish, copious, covering pileus at first and leaving patches on the margin in age or leaving concentric rows of squamules toward the disc. Surface opaque at all times. Context watery gray to brownish, fragile, odor and taste mild; FeSO$_4$ quickly staining base of stipe blackish.

Lamellae broad, close, adnate, vinaceous-buff to avellaneous at first, about "Verona Brown" when mature, edges not beaded or stained.

Stipe 3–5 cm long, 4–6 mm thick, fragile, soon brown in lower half, silky and pallid above, lower half fibrillose or with a slight annular zone.

Spore deposit as air-dried about "Sayal Brown." Spores 7–10 × 5–6 μm, pale clay color in KOH and about the same in Melzer's (not dextrinoid), ellipsoid to ovoid or in profile very obscurely inequilateral, smooth under a high-dry objective.

Hymenium.—Basidia 4-spored, 7–8 μm diam near apex. Pleurocystidia absent. Cheilocystidia a few 26–35 × 10–15 μm and ellipsoid to saccate but mostly ventricose near base and neck becoming elongated and the apex blunt 38–57 × 7–10 × 4–5 μm.

Lamellar and pilear tissues.—Lamellar trama typical of the genus but near the hymenium with a passing flush of red to orange in mounts in Melzer's reagent. Cuticle of pileus a tangled layer of hyphae 3–6 μm broad, yellowish in KOH when fresh, walls often minutely roughened and only weakly refractive in KOH (at most an ixocutis). Hypodermium cellular and rusty brown in KOH (on fresh material rusty ochraceous), walls ± encrusted but no dextrinoid debris observed. Tramal hyphae of pileus typical for the genus (no red content evident in Melzer's).

Habit, habitat, and distribution.—Gregarious under spruce, Elk Camp, Burnt Mt., Pitkin County, Colorado, August 2, 1978 (type, MICH).

Observations.—The dull violaceous shade of the pileus soon vanishes and then the pilei are merely dark brown to blackish. The copious pallid veil is the outstanding field character, but is also found in *H. subcapitatum.* The wide cheilocystidia have been seen in only a few of the other taxa treated here.

33. Hebeloma subcapitatum sp. nov.

Pileus 1–2.5 (3) cm latus, late convexus, expansus vel leviter depressus, subviscidus, cinereocanescens, "Cinnamon-Brown" demum fulvus, squamulosus, glabrescens; odor et gustus submitis. Lamellae pallidae deinde avellaneae, demum obscure cinnamomeae, adnatae, latae, subdistantes. Stipes 1.5–3 cm longus, 4–5 mm crassus, brunnescens, deorsum fibrillosus. Velum copiosum, pallide griseum. Sporae

8.5–10 × 5–6.2 μm, ellipsoideae vel ovoideae, non dextrinoideae, in "KOH" luteolae, subleves. Basidia 23–30 × 6–7 μm, tetraspora. Cheilocystidia 40–65 × 5–8 μm, flexuosa, filamentosa, anguste clavata vel fusoid-ventricosa, subcapitata.

Specimen typicum in Herb. Univ. Mich. conservatum est, Smith 89093; legit prope Independence Pass, Pitkin County, Colorado, 21 Aug 1978, J. Ammirati.

Pileus 1–2.5 (3) cm broad, broadly convex, margin incurved, in age ± expanded plane to slightly depressed, subviscid, with a gray overcast, ground color cinnamon-brown fading to tawny, grayish veil remnants arranged in minute scales, glabrescent, taste and odor not distinctive; FeSO₄ olive-black on discolored part of stipe.

Lamellae broad, close, subdistant, adnate-seceding, pallid then avellaneous, finally dull cinnamon, not beaded or spotted.

Stipe 1.5–3 cm long, 4–5 mm thick, equal, darkening from base up, fibrillose over lower part; veil copious, cortinate, pallid to pale cinereous, not leaving an annular zone.

Spores 8.5–10 × 5–6.2 μm, ovoid to ellipsoid, yellowish in KOH, not dextrinoid, smooth.

Hymenium.—Basidia 4-spored, 25–30 × 6–7 μm, clavate. Pleurocystidia none. Cheilocystidia 40–65 × 5–8 μm, filamentous-capitate, neck flexuous, apex not refractive, some narrowly fusoid-ventricose.

Lamellar and pilear tissues.—Lamellar and pileal tissue (trama) typical for the genus. Cuticle of pileus a thin ixocutis, hyphae 2–3.5 μm diam, yellowish in KOH. Hypodermium cellular, walls encrusted, in KOH yellowish orange to fulvous. Clamp connections present.

Habit, habitat, and distribution.—Gregarious under conifers, Independence Pass area, Pitkin County, Colorado, August 21, 1978 (type, MICH).

Observations.—This species differs from *H. mesophaeum* var. *duplicatum* in having slightly larger spores and a high percentage of elongated filamentous-capitate cheilocystidia. *H. mesophaeum* var. *bifurcatum*, which also belongs in this group, has a hyphoid hypodermium. The diagnostic characters of *H. subcapitatum* are: (1) the velar squamules left on the pileus as the veil breaks, (2) lack of a distinctive odor and taste; (3) predominantly filamentous-capitate cheilocystidia, and (5) a poorly defined ixocutis. *H. subviolaceum* is at once distinguished by having some saccate cheilocystidia.

34. Hebeloma solheimii sp. nov.

Pileus 1.5–4 (6) cm latus, obtusus deinde campanulatus vel plano-umbonatus, subviscidus, ± "Sayal Brown" to "Cinnamon-

Buff," ad marginem fibrillosus vel minute squamulosus, glabrescens; gustus leviter farinaceus, odor mitis. Lamellae pallidae deinde avellaneae demum ± cinnamomeae, confertae, angustae, adnatae. Stipes 2–4 (6) cm longus, 2–5 (8) mm crassus, lacerato-fibrillosus. Velum fibrillosum, subalbidum; stipes deorsum brunnescens. Sporae 8–10 (11) × 5–6.5 μm, subleves, ellipsoideae vel ovoideae vel phaseoliformes, non dextrinoideae. Cheilocystidia 38–66 × 7–11 × 3–5 μm fusoid-ventricosa, rare bifurcata, demum agglutinata.

Specimen typicum in Herb. Univ. Mich. conservatum est, Smith 89471; legit prope Elk Camp, Burnt Mt., Pitkin County, Colorado, 14 Jul 1949.

Pileus 1.5–4 (6) cm broad, convex to obtusely campanulate, surface subviscid, brownish drab to dingy orange-cinnamon to "Sayal Brown" (dull cinnamon) very hoary overall at first from a coating of pallid veil fibrils, in age often covered with minute appressed squamules before becoming glabrous, opaque at all stages. Context watery brown fading to pale buff, taste slight (faintly farinaceous), odor not distinctive; $FeSO_4$ on stipe base instantly blackish green, on the pileus slowly olive.

Lamellae near avellaneous when young, becoming "Sayal Brown," narrow, close, adnate, edges even.

Stipe 2–4 (6) cm long, 2–5 (8) mm thick, equal, rigid, ragged from veil remnants and often with an evanescent annular zone or simply thinly fibrillose, glabrescent, soon bister at the base and the color change progressing upward but apex remaining pallid.

Spore deposit "Clay Color" as air-dried. Spores 8–10 (11) × 5–6.5 μm, appearing smooth in KOH, in Melzer's minutely punctate under oil immersion, not dextrinoid, shape in face view ovate to elliptic, in profile elliptic to ± bean-shaped.

Hymenium.—Basidia 4-spored, 7–9 μm broad, containing numerous hyaline globules as revived in KOH. Pleurocystidia none. Cheilocystidia elongate fusoid-ventricose, 38–66 × 7–11 × 3–5 μm, subacute to obtuse, rarely forked at a short distance below the apex, some present as filaments 3–5 μm diam, hyaline at first but in age agglutinated and the masses rusty brown as revived in KOH or in Melzer's. Gill trama typical of the genus.

Cuticle of pileus a poorly defined layer of hyphae 3–4.5 μm wide with only subgelatinous walls and an outer area of veil hyphae 4–7 μm diam. Hypodermium a layer of rusty brown, wide hyphae appearing cellular in tangential sections, incrustations present but inconspicuous. Pileus trama of ± radial-interwoven hyphae with inflated cells and some dextrinoid debris present. Clamps present.

Habit, habitat, and distribution.—Densely gregarious on an outwash of a stream, Elk Camp, Burnt Mt., Pitkin County, Colorado, July 14, 1979 (type, MICH).

Observations.—The diagnostic features of this species are: (1) the poorly defined ixocutis; (2) the copious pallid veil; (3) the shape of the spores and the fact that they are practically smooth and are not dextrinoid; (4) the agglutinating cheilocystidia as described; (5) the narrow gills, and these grayish when young; and (6) the darkening stipe. It fruited in great abundance in the type locality and the basidiocarps were found to retain their features for periods of up to ten days. Smith 89471 is designated as the type because of the more detailed information available on it. The collection 6235 made by Solheim was from Nash Fork Campground, Albany County, Wyoming at ± 10,000 ft. elevation.

35. Hebeloma mesophaeum (Fr.) Quélet

Champ. Jura et Vosges p. 128. 1872

var. mesophaeum

Pileus 2.5–6 cm broad, at first convex to campanulate, or subumbonate to ± conic, finally expanding to plane or broadly umbonate, at times the central area depressed, viscid, margin grayish brown to near clay color or pinkish buff but the disc usually darker and dull vinaceous-brown (near "Rood's Brown" to "Warm Sepia"), often hoary and with white patches of veil material near the margin or midway to the disc. Context white, thick on the disc, rather thin on margin; odor and taste of radish to not distinctive.

Lamellae broad, narrowly adnate to adnexed or emarginate, at first whitish, finally "Cinnamon-Buff" to "Sayal Brown," close, broad, 3–4 ranks of lamellulae; edges white floccose but not beaded.

Stipe 3–7 cm long, 3–8 mm thick, dingy brownish, darker below especially in age, mealy and pallid near apex, fibrillose-striate below, equal or nearly so; veil white to pallid, fibrillose, typically leaving remnants on or near pileus margin and/or on the stipe, an annular zone often present.

Spore deposit "Dresden Brown" fresh, near "Sayal Brown" as air-dried. Spores 8–10.5 (11) × 5–6 μm, in profile obscurely inequilateral, in face view elliptic to broadly ovate, walls about 0.3 μm thick, minutely marbled under a high-dry lens, pale ochraceous in KOH, not dextrinoid.

Hymenium.—Basidia 4-spored, 24–34 (40) × 7–9 μm. Pleurocystidia none. Cheilocystidia 30–54 (70) × 5–12 × 3–5 μm, subcylindric to elongate fusoid-ventricose, hyaline, wall in neck often flexuous, often with a crook (knee-joint-like) near the base.

Lamellar and pilear tissues.—Lamellar trama typical for the genus (lacking red tints in Melzer's). Cuticle of the pileus a well-developed ixocutis to an ixolattice; hyphae clamped, the walls refractive and the hyphae 1.5–4 μm diam. Hypodermium cellular, the walls dark brown in KOH. Tramal hyphae of the pileus lacking a red reaction with Melzer's.

Habit, habitat, and distribution.—Solitary to gregarious or clustered on soil under conifers, especially spruce, late fall, summer, and early spring. It is widely distributed throughout the western United States. A typical specimen is Smith 49417.

Observations.—This is still a collective species as described here for North American material. There is little likelihood of clearing up the problems of its variation in Europe until a type collection consistent with the protologue of the species is established for it. The problems rest mainly with variation in a number of characters important in the subgenus. The spores are borderline between the bean-shaped and inequilateral types. We as yet do not have much data on the iodine reactions of the spores and tissues of the basidiocarp on European material, and the same is true for the color of the veil and the changes it undergoes as the stipe tissues become discolored. Nearly all of the following varieties were carried as "species" in our preliminary manuscripts.

Bruchet's (1970) account covers our "typical" material rather well. The important combination of characters as we have found it for North American collections is as follows: Either the taste or odor is raphanoid to some extent, and the taste soon becomes bitterish in most collections, the pileus is dark brown when moist, and the stipe darkens at the base and the change progresses upward, the veil is white, the spores are 8–10 (11) × 5–6 μm and not truly inequilateral but ± bean-shaped in profile view. Many cheilocystidia are ventricose at or near the base, and in optical sections the wall of the neck is seen to be wavy to irregular. Finally, the apex of the cheilocystidium typically is subacute to obtuse (not significantly enlarged).

KEY TO VARIANTS

5. Veil pallid (not appreciably discoloring). ....
    .... 35 var. *mesophaeum* and var. *longipes* (h) (p. 170)
5. Veil buff colored in young basidiocarps .... 35d. var. *imitatum*
    6. Veil pallid but soon lutescent on the darkened area of the stipe
    .... 35e. var. *aspenicola*
    6. Not as above 7
7. Pileus ± chestnut brown at first; odor and taste ± pungent   .. 35f. var. *castaneum*
7. Not as above 8
    8. At maturity some cheilocystidia spathulate or forked at the apex
    .... 35g. var. *bifurcatum*
    8. Not as above 9
9. Spores 7–9 (10) × 5–5.5 μm .... 35h. var. *duplicatum*
9. Spores 8–10 × 6–7.5 μm 10
    10. Veil grayish; pileus squamulose .... 35i. var. *insipidum*
    10. Not as above 11
11. Veil pallid; taste farinaceous .... 35j. var. *fluviatile*
11. Veil buff colored .... 35k. var. *similissimum*

35a. Hebeloma mesophaeum var. lateritium
(Murrill) comb. nov.

Hebeloma lateritium Murrill, North Amer. Flora 10: 224. 1917.

Pileus 5 cm broad, convex, becoming nearly plane with age, umbonate, surface distinctly viscid, smooth, glabrous, lateritius, margin entire, not striate, avellaneous or, as moisture escapes, cream color.

Lamellae sinuate, rather broad and ventricose, crowded, pallid to clay color, conspicuously whitish pubescent on the edges.

Stipe 6 cm long, 7–10 mm thick, enlarged below, white (as described by Murrill), conspicuously fibrillose, fleshy.

Spores 8–11 × 4–5.5 μm, narrowly ovate in face view, obscurely bean-shaped to elliptic in profile view, very pale clay color in KOH, appearing smooth under a 4-mm high-dry objective, walls thickened somewhat but no apical pore evident, not dextrinoid.

Hymenium.—Basidia both 2- and 4-spored, 26–30 × 6–7 μm. Pleurocystidia none. Cheilocystidia 30–73 × 7–11 × 4–6 μm, fusoid ventricose or some filamentous, neck usually straight, wall hyaline to yellowish in KOH; not agglutinating and few seen with a slight apical enlargement.

Lamellar and pilear tissues.—Lamellar trama typical for the genus. Cuticle of pileus an ixolattice of hyaline hyphae 3–5 μm diam; clamp connections present. Pileal trama typical for the genus. Hypodermium a distinct layer of brown hyphae apparently cellular but the cells reviving poorly. Degree to which incrustations were present on the elements not established.

Habit, habitat, and distribution.—Gregarious on sandy soil at the edge of a forest, Seattle, Washington, October 20–November 1, 1911, Murrill 295 (NY). Apparently known only from the type collection.

Observations.—The color of the dried cap is about like that of *Naematoloma sublateritium* (Fr.) Karsten at least on the disc. As dried, the stipe appears to be the type that darkens from the base up in age, though this is obscured by the white fibrils of the veil. Murrill described it simply as white. For *H. mesophaeum* he described the stipe as "becoming rusty"—which shows he was aware of this feature in this group. The distinctive combination of features shown by Murrill's type is: the brick red disc of the pileus; the thick stipe not discoloring markedly, and a copious white veil. The ixocutis of the pileus is better developed than in var. *mesophaeum,* but, as in most of the distinguishing characters, the difference is one of degree. For this reason we reduce the species to varietal status under *H. mesophaeum.*

35b. Hebeloma mesophaeum var. velovinaceum var. nov.

Pileus 2–3.5 cm latus, late convexus, glaber, subviscidus, griseovinaceus, odor et gustus mitis. Lamellae "Vinaceous-Buff" demum ± cinnamomeae, latae, subdistantes. Stipes 3–4 cm longus, 3–4 mm crassus, deorsum attenuatus. Velum "Vinaceous-Buff." Sporae 8–11 × 5–6.5 μm, ovoideae vel ellipsoideae, non dextrinoideae. Cheilocystidia 38–64 × 6–11 μm, filamentosa, subcapitata vel fusoide ventricosa.

Specimen typicum in Herb. Univ. Mich. conservatum est, Smith 51374; legit prope Lizard Head Pass, San Juan County, Colorado, 27 Jul 1956.

Pileus 2–3.5 cm broad convex becoming plane, margin incurved at first, glabrous, subviscid but soon dry, opaque, color a grayish vinaceous-brown (in the herbarium ± "Sayal Brown"). Context thin, brownish, odor and taste mild.

Lamellae broad, subdistant, depressed adnate, vinaceous-buff becoming ± cinnamon, edges neither beaded nor spotted.

Stipe 3–4 cm long, 3–4 mm thick near apex, equal or narrowed downward, bister at base; surface fibrillose from veil remnants which are pale vinaceous (pinkish gray), no annular veil-line evident.

Spores 8–11 × 5–6.5 μm, ovoid to ellipsoid, wall thin and appearing smooth, pale ochraceous in KOH singly, pale rusty brown in Melzer's but soon paler on standing (not dextrinoid).

Hymenium.—Basidia 4-spored, 7–10 μm wide. Pleurocystidia none. Cheilocystidia elongate-clavate to narrowly fusoid-ventricose, 38–64 × 6–11 μm or ± cylindric and up to 80 × 5 μm, the apex obtuse to subcapitate, hyaline, somewhat refractive.

Lamellar and pilear tissues.—Cuticle of pileus an ixolattice (very easily removed), hyphae 2–4 μm diam, hyaline, ± refractive, clamps present. Hypodermium hyphoid, ± tawny in KOH, redder in Melzer's, some dextrinoid debris present.

Habit, habitat, and distribution.—Scattered to gregarious under spruce, Lizard Head Pass, San Juan County, Colorado, July 27, 1956 (type, MICH).

Observations.—The combination öf vinaceous-buff veil (pinkish gray), and lack of any distinctive odor and taste distinguish this variety from var. *mesophaeum* as admitted here, but a close relationship between them is obvious. The most useful difference, in studying herbarium material, is the hyphoid hypodermium but the difference between cellular and hyphoid can be bothersome. In the course of its development, the layer may be hyphoid first and ± cellular at the end point of development.

35c. Hebeloma mesophaeum var. subobscurum var. nov.

Pileus 3–5 cm latus, obtusus demum planus vel obtuse umbonatus, leviter fibrillosus, glabrescens, rufobrunneus, ad marginem saepe brunneogriseus. Contextus pallidus; odor et gustus leviter raphanoideus. Lamellae adnexae, "Vinaceous-Buff" juvenis, demum cinnamomeae, confertae, latae. Stipes 4–6 cm longus, 3–8 mm crassus, pallidofibrillosus, deorsum obscurus, aequalis, siccus. Velum fibrillosum, pallidum, evanescens. Sporae 8–10 (11) × 5–6 μm, ± ellipsoideae, subleves, non dextrinoideae. Cheilocystidia 34–60 × 4–8 μm, plerumque fusoide ventricosa.

Specimen typicum in Herb. Univ. Mich. conservatum est, Smith 46587; legit prope, Seven-Devils Mts., Riggins, Idaho, 23 Aug 1954.

Pileus 3–5 cm broad, obtuse, expanding to obtusely umbonate to plane, when young "Benzo Brown" to "Cinnamon Drab" and coated with a thin layer of fibrils, glabrescent and color then "Verona Brown" to "Warm Sepia" (reddish to rusty brown), viscid to subviscid, soon appearing dry. Context pallid, odor and taste slightly raphanoid.

Lamellae adnexed, "Vinaceous-Buff" (grayish pink) when young, gradually grayer ("Avellaneous"), finally dingy cinnamon from the spores, close, broad, edges uneven, not agglutinating.

Stipe 4–6 cm long, 3–8 mm thick, pallid from a thin fibrillose coating, darkening from the base up in age to a sordid brown ("Bister" or darker), tubular, equal. Veil fibrillose, pallid, thin, evanescent (scarcely leaving a zone on the stipe.

Spores 8–10 (11) × 5–6 μm, subelliptic in profile, elliptic in face view, wall thin, appearing smooth, merely yellowish in KOH, not dextrinoid.

Hymenium.—Basidia 2- to 4-spored, 27–32 × 6–8 μm, ± clavate. Pleurocystidia none. Cheilocystidia 34–60 × 3–5 × 4–12 μm, subcylindric to fusoid-ventricose.

Lamellar and pilear tissues.—Lamellar trama of slightly interwoven hyphae 4–10 μm diam. Pileus trama of hyphae ± radially arranged. Cuticle of pileus a thin ixocutis of hyphae 1.5–3 μm diam, clamp connections present. Hypodermium of brown cells (mounted in KOH).

Habit, habitat, and distribution.—On soil in coniferous woods, Seven-Devils Mts., Idaho (near Riggins), August 23, 1954 (type, MICH).

Observations.—When fresh the pileus appeared dull and was dry to the touch. Under the microscope, however, it is clear that gelatinization of the hyphae of the cuticle is present to some extent.

35d. Hebeloma mesophaeum var. imitatum var. nov.

Pileus 2–3 cm latus, obtusus demum plano-umbonatus, subviscidus, tenuiter fibrillosus, glabrescens, obscure cinnamomeus; odor et sapor raphaninus. Lamellae latae, confertae adnatae, obscure cinnamomeae. Stipes ± 3 cm longus, ± 4 mm crassus, aequalis, brunnescens. Velum pallide argillaceum. Sporae 8–10 × 5–6 μm, ovoideae vel ellipsoideae vel phaseoliformes, non dextrinoideae. Cheilocystidia fusoid-ventricosa, 37–65 × 5–11 × 3–6 μm.

Specimen typicum in Herb. Univ. Mich. conservatum est, Smith 53134; legit prope Trout Lake, San Juan County, Colorado, 24 Jul 1956.

Pileus 2–3 cm broad, obtuse expanding to plano-umbonate, surface subviscid beneath a thin coating of buff veil fibrils, ± glabrescent, ground color dull cinnamon ("Sayal Brown"), odor and taste raphanoid.

Lamellae broad, close, adnate, dull brownish becoming about "Sayal Brown."

Stipe ± 3 cm long, ± 4 mm thick, equal; fragile, darkening to dark brown from the base upward; veil fibrils buff colored (but no young specimens were available).

Spores 8–10 × 5–6 μm, ovate to elliptic in face view, oblong to bean-shaped in profile (apex more rounded than usual), not dextrinoid (merely yellowish in Melzer's), smooth or nearly so.

Hymenium.—Basidia 30–32 × 7–9 μm, 4-spored. Pleurocystidia absent. Cheilocystidia fusoid-ventricose, in widest part 5–11 μm, becoming greatly elongated in age (± 80 μm or more), hyaline and thin-walled.

Lamellar and pilear tissues.—Lamellar trama typical for the genus. Cuticle of pileus a thin ixocutis, the hyphae 2–7 μm diam, hyaline, clamped. Hypodermium intermediate in type, the hyphae 3–17 μm diam, rusty brown and walls incrusted (in KOH). Hyphae of pileus trama typical of the genus (no red tints in Melzer's).

Habit, habitat, and distribution.—Scattered to gregarious under mixed conifer-aspen stands, Trout Lake, San Juan County, Colorado, July 24, 1956 (type, MICH).

Observations.—The buff-colored veil and the tendency of the cheilocystidia to elongate late in the development of the basidiocarp are the characters we focused upon in this variety.

35e. Hebeloma mesophaeum var. aspenicola var. nov.

Illus. Figs. 9–10.

Pileus 1.5–2 cm latus, obtuse conicus deinde plano-umbonatus, ad marginem fibrillosus, ad centrum glaber, viscidus, nitens, rubrobrunneus; odor et gustus pungens. Lamellae confertae, latae, adnatae, obscure cinnamomeae. Stipes 3–5 cm longus, 3–5 mm crassus, fibrillosus, brunnescens. Velum album demum subochraceum. Sporae 7–10 × 5–6 μm, subleves, ovoideae vel ellipsoideae vel phasaeoliformes, non dextrinoideae. Cheilocystidia (35) 40–62 × 6–11 × 3–5 μm, fusoidventricosa.

Specimen typicum in Herb. Univ. Mich. conservatum est, Smith 90192; legit prope Snowmass Village, Pitkin County, Colorado, 2 Sep 1979.

Pileus 1.5–2 cm broad, obtusely conic becoming plane or expanded umbonate, margin bent in somewhat and very sparsely fibrillose-fringed from fibrils of a thin pallid veil; surface glabrous, viscid, shining when dry (in situ); color "Verona Brown" or redder, becoming "Clay Color" finally overall, margin at times pale pinkish buff. Context whitish when faded, firm, odor faintly pungent, taste ± pungent (not raphanoid), KOH on cuticle brownish, FeSO₄ olive-gray on base of stipe.

Lamellae broad, close, adnate, pale brown then dull cinnamon, not beaded or spotted, edges even.

Stipe 3–5 cm long, 3–5 mm thick, equal, whitish, thinly fibrillose from a pallid veil which often leaves a faint superior zone, veil fibrils discolor to ± clay color as basal area of stipe darkens beneath them.

Spores 7–10 × 5–6 μm, ovate to elliptic in face view, in profile ± bean-shaped to ovate, smooth in KOH but in Melzer's slightly punctate, not dextrinoid (yellowish in Melzer's).

Hymenium.—Basidia 4-spored, 8–10 μm broad, clavate, with hyaline globules as revived in KOH. Pleurocystidia none. Cheilocystidia (33) 40–62 × (6) 7–11 μm, neck 3–5 μm diam, fusoid-ventricose, not agglutinating.

Lamellar and pilear tissues.—Lamellar trama in Melzer's showing a flush of red to reddish orange along the base of the hymenium. Cuticle of pileus a well-defined ixolattice of hyphae 2–4 μm diam and with ± gelatinized walls (some nongelatinized veil hyphae at times overlaying this layer). Hypodermium cellular, the cells somewhat enlarged and pale rusty brown in KOH, incrustations (or wall thickenings) present but not conspicuous; copious dextrinoid debris present near and in the layer. Tramal hyphae of pileus ochraceous-hyaline in KOH, smooth, radial-interwoven, cells inflated to 12–30 μm diam, narrower near the gill edge. Clamp connections present.

Habit, habitat, and distribution.—Gregarious under aspen, Sam's Knob, Snowmass Village, Pitkin County, Colorado, September 2, 1979 (type, MICH).

Observations.—In radial sections the hypoderm is seen to consist of short to long inflated hyphal cells. In KOH the color is at first pronounced but fades considerably in 15 minutes. Dextrinoid debris is present in the mount variously: in the gill or pileus trama, apex of the stipe, or occasionally in the ixolattice. The shiny reddish brown pileus, the weak FeSO₄ reaction on the base of the stipe and the habitat under aspen distinguish this variety. *H. testaceum* in the sense of Bruchet has amygdaliform spores (8) 10–12 (13) × 5–6 (6.5) μm.

35f. Hebeloma mesophaeum var. castaneum var. nov.

Pileus 1–2.5 (3) cm latus, obtuse conicus vel subcampanulatus, tenuiter fibrillosus deinde glaber, ad centrum ± castaneus demum subargillaceus; odor et gustus subraphaninus. Lamellae pallidae demum cinnamomeae, latae, confertae adnatae. Stipes 3–4 cm longus, 3–4 mm crassus, griseofibrillosus, brunnescens. Sporae 8.5–10 × 5–6 μm, ovoideae vel ellipsoideae, non dextrinoideae, subleves, in "KOH" ochraceae. Cheilocystidia anguste fusoid-ventricosa, 45–65 × 5–9 × 3–4 μm.

Specimen typicum in Herb. Univ. Mich. conservatum est, Smith 88922; legit prope Elk Wallow, Pitkin County, Colorado, 9 Aug 1978, Thiers and Smith.

Pileus 1–2.5 (3) cm broad, obtusely conic expanding to broadly conic to convex, surface glabrous over disc, thinly coated with veil remnants on the margin, at times with inconspicuous patches; disc chestnut brown to cinnamon-brown, margin paler and grayer, some with a faint ochraceous flush. Context thin, watery brown, pale dingy buff faded; odor and taste pungent (± raphanoid ?) FeSO$_4$ blackish at base of stipe.

Lamellae broad, close adnate, pallid then dull cinnamon, not beaded and cheilocystidia not agglutinated.

Stipe 3–4 cm long, 3–4 mm thick, equal, soon dark brown below, upper half grayish fibrillose from the copious veil, veil-line soon evanescent.

Spores 8.5–10 × 5–6 μm, ovoid to ellipsoid, ochraceous in KOH, not dextrinoid, very weakly ornamented (subleves).

Hymenium.—Basidia 4-spored, 22–25 × 5–6 μm. Pleurocystidia none. Cheilocystidia narrowly fusoid-ventricose, 45–65 × 5–9 × 3–4 μm.

Lamellar and pilear tissues.—Lamellar trama typical for genus. Cuticle of pileus a thin ixocutis, hyphae 2–4 μm diam, yellowish in KOH. Clamps present. Hypodermium hyphoid to intermediate, orange-brown in KOH.

Habit, habitat, and distribution.—Gregarious on conifer duff, Elk Wallow, Fryingpan River, Pitkin County, Colorado, August 9, 1978 (type, MICH).

Observations.—In the fresh state this variety features a chestnut brown pileus thinly coated with veil fibrils which soon vanish save for an area along the margin. The veil is grayish and in KOH the spores are ochraceous. The hypodermium is intermediate between cellular and hyphoid, and the odor and taste though not mild are not truly raphanoid either. Spruce dominated the habitat.

35g. Hebeloma mesophaeum var. bifurcatum var. nov.

Hebeloma mesophaeum var. *mesophaeum* subsimile sed cheilocystidia subcylindrica vel flexuosa, ad apicem obtusa, spathulata vel rare bifurcata. Hypodermium intermedia.

Specimen typicum in Herb. Univ. Mich. conservatum est, Smith 88816, legit prope Independence Pass, Pitkin County, Colorado, 31 Jul 1978.

Pileus 1.5–3.5 cm broad, obtuse, expanding to ± conic-campanulate, surface hoary from a thin coating of veil fibrils, glabrescent, "Cinnamon-Brown" on disc, paler on margin, surface becoming rimose to areolate; odor and taste mild; FeSO$_4$ quickly olive-fuscous on base of the stipe.

Lamellae broad and ventricose, subdistant, adnate then adnexed, avellaneous, becoming ± cinnamon, not beaded.

Stipe 2–2.5 cm long, 4–5 mm thick, soon darkening below, hollow in age, surface ± fibrillose from a grayish veil, near avellaneous above and near the apex silky, brunnescent as it ages.

Spores 8.5–10.5 × 5–6 µm, appearing smooth to faintly rugulose, ellipsoid, pale yellow in KOH, not dextrinoid, in profile view ± elliptic.

Hymenium.—Basidia 4-spored. Pleurocystidia none. Cheilocystidia 45–65 × 7–9 µm, some with an enlargement near the base, apex obtuse to spathulate or rarely forked or branched.

Lamellar and pilear tissues.—Lamellar trama typical for the genus. Cuticle of pileus a poorly defined ixocutis, the hyphae ochraceous in KOH and finely ornamented. Hypodermium hyphoid for the most part, rusty brown in KOH and incrusted pigment on the hyphal walls, hyphae up to 20 µm diam. Clamp connections present.

Habit, habitat, and distribution.—Gregarious on needle carpet under pine, near Independence Pass (Lost Man Camp), Pitkin County, Colorado, July 31, 1978 (type, MICH).

Observations.—The spathulate to forked cheilocystidia are most readily observed in mature to old pilei, and the same is true for the rimose feature of the pileus. *H. subrimosum* has slightly larger spores and a thicker stipe.

35h. Hebeloma mesophaeum var. duplicatum var. nov.

Pileus 1–3 cm latus, obtuse umbonatus, griseofibrillosus, glabrescens, triste cinnamomeus; sapor subpungens. Lamellae griseobrunneae demum subcinnamomeae, confertae, latae. Stipes 3–5 cm longus, 3–5 mm crassus, deorsum brunnescens, sursum pallidus, leviter fibrillosus. Velum griseum vel ± pallidum, sublutescens. Sporae 7–9 (10) × 5–5.5 µm, non dextrinoideae, ellipticae vel ovatae. Cheilocystidia 36–52 × 4–8 µm, subcylindrica, leviter ventricosa vel anguste clavata. Cuticula pileorum ixocutis est.

Specimen typicum in Herb. Univ. Mich. conservatum est, Smith 90058A; legit prope Elk Wallow, Pitkin County, Colorado, 29 Aug 1979.

Pileus 1–3 cm broad, obtuse becoming broadly umbonate, coated with a thin layer of grayish veil fibrils (heaviest along the margin), ground color ± dark cinnamon (± "Sayal Brown") when moist, paler when faded (about "Cinnamon-Buff"). Context with a slight pungent odor and taste (subraphanoid); KOH on cuticle of pileus brownish; FeSO$_4$ on stipe base dark grayish brown.

Lamellae broad and ventricose, close, adnate to adnexed, avellaneous to wood brown, finally about "Sayal Brown" (dull cinnamon), very slightly beaded.

Stipe 3–5 cm long, 3–5 mm thick, ± equal, brownish within

especially below, surface pallid from a thin but distinct fibrillose veil, the latter breaking up into zones and patches and these becoming darker as the context darkens.

Spores 7–9 (10) × 5–5.5 μm, smooth (?), not dextrinoid, in face view ovate to elliptic, in profile ± ovate, apex rounded, wall only faintly brownish in KOH.

Hymenium.—Basidia 4-spored. Pleurocystidia none. Cheilocystidia 36–52 × 4–8 μm, cylindric, narrowly clavate, or weakly ventricose near the base, evenly distributed, walls thin and hyaline as revived in KOH.

Lamellar and pilear tissues.—Lamellar trama typical for the genus. Cuticle of pileus an ixocutis. Hypodermium of the intermediate type, yellow in KOH when fresh. Context of pileus typical of the genus (not red in Melzer's).

Habit, habitat, and distribution.—Gregarious on outwash area of a stream under lodgepole pine and spruce mixed, Fryingpan River, Pitkin County, Colorado, near Elk Wallow, August 29, 1979 (type, MICH).

Observations.—The spores of this variety are smaller than in var. *mesophaeum,* the odor and taste are doubtfully raphanoid, and the $FeSO_4$ reaction on the base of the stipe was merely grayish brown, not olive or olive-black.

35i. Hebeloma mesophaeum var. insipidum var. nov.

A typo differt: odor et gustus subpungens; cheilocystidia (46) 57–86 × 7–15 × 3.5–5 μm, fusoid-ventricosa; pileus fibrilloso-squamulosus; velum cinereum.

Specimen typicum in Herb. Univ. Mich. conservatum est, Smith 88791; legit prope Elk Wallow, Fryingpan River, Pitkin County, Colorado, 30 Jul 1978.

Pileus 1.5–4 cm broad, obtusely conic becoming broadly conic to expanded-umbonate, margin incurved, surface hygrophanous but opaque, when young "Wood Brown" (brownish gray) and not viscid, soon "Dresden Brown" to "Snuff Brown" (dull date brown) and slowly fading to alutaceous, surface at first finely fibrillose-squamulose, margin with grayish patches from the fibrillose veil. Context watery brown, fragile, odor and taste faintly pungent (not clearly distinctive); $FeSO_4$ quickly staining base of stipe dark olive.

Lamellae moderately broad, close, adnate, grayish at first, soon pale brown and finally dark rusty brown (about concolorous with pileus), edges neither spotted nor beaded.

Stipe 3–4.5 cm long, 3–4 mm thick, equal, soon hollow, fragile, pallid at first but soon darkening to bister from the base up, somewhat fibrillose from the grayish veil but no annular zone evident when it breaks.

Spore deposit as air-dried about "Sayal Brown"; spores (8) 9–11 × 5–6 μm, pale clay color in KOH, paler in Melzer's (not dextrinoid), obscurely mottled (as seen mounted in Melzer's), somewhat bean-shaped to obscurely inequilateral in profile, elliptic to ovate in face view, apex obtuse.

Hymenium.—Basidia 4-spored, 8–9 μm broad. Pleurocystidia none. Cheilocystidia abundant (46) 57–86 × 7–15 × 3.5–5 μm, fusoid-ventricose, the neck slightly tapered to a subacute to obtuse apex, flexuous; hyaline, smooth, wall not appreciably refractive.

Lamellar and pilear tissues.—Lamellar trama typical for the genus. Cuticle of pileus a poorly defined ixocutis, the hyphae 2–3.5 μm wide, only slightly gelatinous, clamped, smooth to slightly encrusted. Hypodermium "cellular" (in tangential sections), cell walls in KOH rusty brown (about the same in Melzer's), incrustations and/or wall thickenings numerous. Tramal hyphae typical of the genus (not with red content as mounted in Melzer's).

Habit, habitat, and distribution.—Densely gregarious on soil, Fryingpan River, Pitkin County, Colorado, above Elk Wallow, July 30, 1978 (type, MICH).

Observations.—The grayish veil and a ± distinctly pungent taste and odor distinguish this variety from var. *mesophaeum.* Other differences such as the squamulose pileus margin at first, and an appreciable number of very large cheilocystidia along with a larger number of spores bean-shaped in profile view, will be noted.

35j. Hebeloma mesophaeum var. fluviatile var. nov.

Pileus 1.5–4 cm latus convexus demum campanulatus, subviscidus, griseobrunneus demum sordide cinnamomeus, canescens demum nudus. Odor debilis (± raphaninus). Lamellae confertae, angustae, adnatae, avellaneae demum sordide cinnamomeae. Stipes 2–4 (6) cm longus, 2–5 mm crassus, strictius, fibrillosus, deorsum demum triste spadiceus. Sporae 8–10.5 × 5–6 μm, non dextrinoideae. Cheilocystidia 38–54 × 7–11 × 3–5 μm, fusoid-ventricosa. Cuticula pileorum subixocutis est. Hypodermium in "KOH" fulvobrunneum.

Specimen typicum in Herb. Univ. Mich. conservatum est, Smith 89471; legit prope Elk Camp, Burnt Mt., Pitkin County, Colorado, 14 Jul 1979.

Pileus 1.5–4 cm broad, convex to obtusely campanulate, surface subviscid, brownish drab to dingy orange-cinnamon (near "Sayal Brown," grayish brown becoming dull orange-cinnamon), very hoary overall at first from a coating of pallid veil fibrils, in age often covered with minute appressed squamules before becoming glabrous, opaque at all stages. Context watery brown fading to pale buff, taste slightly (faintly) farinaceous, odor not distinctive (weakly pungent to ± raphanoid when context is crushed), $FeSO_4$ on base of stipe instantly blackish green, on cuticle of pileus slowly olive.

Lamellae narrow, close, adnate, near avellaneous when young, becoming dull cinnamon at maturity, edges even, neither beaded nor spotted.

Stipe 2–4 (6) cm long, 2–5 mm thick, equal, rigid, ragged from veil remnants and often with an evanescent annular zone, or simply thinly fibrillose, glabrescent, soon bister at the base and the color change progressing upward but apex remaining pallid.

Spore deposit "Clay Color" as air-dried. Spores 8–10.5 × 5–6 μm, appearing smooth in KOH (under a high-dry lens), in Melzer's nondextrinoid and minutely punctate under an oil-immersion lens; shape in face view ovate to elliptic, in profile elliptic to ± bean-shaped or rarely obscurely inequilateral.

Hymenium.—Basidia 4-spored, 7–9 μm broad near apex, containing numerous hyaline globules as revived in KOH. Pleurocystidia none. Cheilocystidia elongate fusoid-ventricose, 38–54 × 7–11 × 3–5 μm, subacute to obtuse, rarely forked at a short distance below the apex, some present as filaments 3–5 μm diam; hyaline at first but in age agglutinated and the masses rusty brown as revived in KOH or in Melzer's.

Lamellar and pilear tissues.—Lamellar trama typical of the genus. Cuticle of pileus a poorly defined layer of hyphae 3–4.5 μm wide and with only subgelatinous walls (an outer layer of veil hyphae 4–7 μm wide at times observed over the cuticle). Hypodermium a layer of rusty brown wide hyphae hyphoid to cellular, incrustations present but inconspicuous. Pileus trama of ± radial interwoven hyphae with inflated cells and some dextrinoid debris present. Clamps present.

Habit, habitat, and distribution.—Densely gregarious on an outwash of a mountain stream, Elk Camp, Burnt Mt., Pitkin County, Colorado, July 14, 1979 (type, MICH).

Observations.—The important taxonomic characters are: the narrow lamellae, farinaceous taste, squamulose pileus, nondextrinoid spores, agglutinating cheilocystidia, the darkening stipe, and the *H. mesophaeum*-like spores. We describe it at the varietal level because the taste is weak and will not be detectable by some people, the squamules on the pileus soon vanish and besides these are merely veil remnants and all the related species also have veils of ± similar texture.

35k. Hebeloma mesophaeum var. similissimum var. nov.

Pileus 2–4 cm latus, plano-unbonatus, viscidus, glaber, fulvus sed ad marginem griseobrunneus et fibrillosus. Odor et gustus pungens (subraphaninus). Lamellae confertae, adnatae, obscure cinnamomeae. Stipes 3–4 cm longus, 3–4 mm crassus, brunneus. Velum subargillaceum. Sporae 8.5–10 (11) × 5.5–6.8 μm, oblongae vel ovoideae, subleves. Cheilocystidia (35) 42–67 × 4–10 × 3–5 μm, fusoid-ventricosa, obtusa.

Specimen typicum in Herb. Univ. Mich. conservatum est, Smith 87091; legit prope Independence Pass, Pitkin County, Colorado, 5 Aug 1976.

Pileus 2–4 cm broad, plano-umbonate, viscid, glabrous, tawny over the disc, margin paler and grayish, often with thin clay color patches of veil fibrils along the margin (as dried). Context white, odor and taste pungent, $FeSO_4$ on base of stipe merely brownish.

Lamellae close, adnate, narrow, pinkish brown when young, ± "Verona Brown" in age (a dull reddish brown), not beaded.

Stipe 3–4 cm long, 3–4 mm thick, equal, solid, pallid within, brownish at surface, with an annular fibrillose zone from the broken veil, pallid and scurfy near the apex.

Spores 8.5–10 (11) × 5.5–6 (6.8) μm, ellipsoid to ovoid, in profile some of them oblong, ± smooth under a high-dry objective, not dextrinoid, wall thin.

Hymenium.—Basidia 4-spored. Pleurocystidia none. Cheilocystidia (35) 42–67 × 4–10 × 3–5 μm, fusoid-ventricose, apex obtuse to slightly enlarged, often with a wavy neck.

Lamellar and pilear tissues.—Cuticle of pileus an ixolattice fairly well developed, the hyphae 2–7 μm diam, branched, clamped at the septa. Hypodermium cellular, reddish brown in KOH. Pileal trama of radial hyphae 3–15 μm diam, some hyphae with inflated cells, ochraceous in KOH.

Habit, habitat, and distribution.—Scattered under conifers, Independence Pass area, Pitkin County, Colorado, August 5, 1976 (type, MICH).

Observations.—The narrow gills, color of the veil, and brownish instead of olive-black $FeSO_4$ reaction on the base of the stipe distinguish this variant from our concept of *H. mesophaeum* var. *mesophaeum*.

Section HEBELOMA

Subsection PALLIDAE subsect. nov.

Sporae ± inequilaterales; pileus albidus vel pallidus, saepe tarde subochraceus.

Typus: *Hebeloma barrowsii*

Spores inequilateral; pileus white to pallid, becoming cream colored.

KEY TO SPECIES

1. Stipe at apex 2–3 cm thick; taste farinaceous .... 36. *H. farinaceum*
1. Stipe 4–10 mm thick; taste not as above 2
   2. Stipe floccose-squamulose overall; odor and taste not distinctive
   .... 37. *H. kanousiae*
   2. Stipe thinly fibrillose; odor and/or taste distinctive 3
3. Stipe becoming clay color from base upward; odor and taste raphanoid
   .... 38. *H. salmonense*
3. Stipe scarcely discoloring; odor "woody" (Barrows); taste unpleasant
   .... 39. *H. barrowsii*

36. Hebeloma farinaceum Murrill

North Amer. Flora 10: 226. 1917

Pileus 8–10 cm broad, convex to nearly plane, not umbonate, cream colored to light buff, glabrous, viscid, margin pallid, entire. Context thick, white, odor and taste farinaceous.

Lamellae sinuate, yellowish to subfulvous, rather narrow, plane or arcuate, crowded, edges entire and concolorous with faces.

Stipe 6–8 cm long, 20–30 mm thick, white, enlarged below, fleshy, solid. Veil fibrillose, evanescent, the remains decorating the stipe.

Spores in deposit "Cinnamon-Brown" (after sixty years in storage), brownish with a yellow tint in 2 percent KOH, 10–12 (13) × 6–7.5 μm, inequilateral in profile, fusoid in face view, rugulose to rugose, wall ± 0.3 μm thick, apiculus relatively short.

Hymenium.—Basidia 35–42 × 6–7 μm, 4-spored. Pleurocystidia and cheilocystidia none.

Lamellar and pilear tissues.—Lamellar trama of subparallel hyphae. Cuticle of pileus an ixocutis, the hyphae partly gelatinized. Hypodermium of dingy nonpigmented poorly defined hyphae. Clamp connections not observed (but the material was unsuitable for an accurate study).

Habit, habitat, and distribution.—On soil, under oaks, near Stanford University, California, collected by James McMurphy, January 11, 1912, Murrill 126 (type, NY).

Observations.—The failure to find clamp connections could easily be due to the condition of the material and so the feature is not emphasized here. The sharp odor, thick stipe, and distinctly roughened spores under the light microscope are distinctive. The species appears to be related to *H. fastible,* but further studies from ample material of both are needed to characterize them properly.

37. Hebeloma kanousiae sp. nov.

Pileus 2–4 cm latus, albus vel lacteus, leviter viscidus, glaber. Contextus albus, odor et gustus mitis. Lamellae albidae demum "Clay Color," angustae, confertae. Stipes 3–5 cm longus, 4–7 mm crassus, albissimus, squamas albas floccoso-farinaceas gerens, siccus. Sporae 12–15 × 6–8 μm, sublimoniformes, non dextrinoideae. Cheilocystidia 30–60 × 5–8 μm, cylindrica, ad basin ± ventricosa.

Specimen typicum in Herb. Univ. Mich. conservatum est; legit prope Medicine Bow Mts., Wyoming, 28 Aug 1923; B. B. Kanouse.

Pileus 2–4 cm broad, convex, margin at first incurved, expanding to nearly plane or remaining broadly convex, often somewhat irregular in age, surface glabrous, not shining (merely subviscid), margin at first "pulverulent flocculose-appressed" (Kauffman), color white to milk white or cream color overall; odor and taste slight (not distinctive).

Lamellae narrow, adnate or arcuate-subdecurrent, close, thin, whitish becoming "Light Pinkish Cinnamon" and finally "Clay Color," edges beautifully serrulate when young, white fimbriate to somewhat eroded in age.

Stipe 3–5 cm long, 4–7 mm thick, at first stuffed by a white pith then hollowed, equal, scarcely bulbous at the base, shining white, "covered throughout by floccose-mealy scales" (Kauffman), surface beneath the scales soft and fibrous.

Spores 12–15 × 6–8 μm, distinctly roughened, pale clay color in KOH; shape in profile inequilateral, in face view ovate and the apex tending to be snoutlike, merely ochraceous in Melzer's (not dextrinoid).

Hymenium.—Basidia 34–40 × 7–8 μm, 4-spored. Pleurocystidia none. Cheilocystidia 35–60 × 5–8 μm, filamentous to elongate-clavate, hyaline, smooth, thin-walled.

Lamellar and pilear tissues.—Lamellar trama typical for the genus. Subhymenium not distinctively colored in Melzer's. Cuticle of pileus an ixocutis, the hyphae 2–3 μm diam, interwoven, appressed. Hypodermium hyphoid and not distinctly colored in KOH or Melzer's, the layer merely more compact than the trama beneath it. Clamp connections present.

Habit, habitat, and distribution.—On leaf mold under aspen near the edge of a swamp; collected by B. B. Kanouse, Medicine Bow Mts., Wyoming (above Centennial), August 28, 1923 (type, MICH).

Observations.—The large spores, white to pale or dark cream-colored pileus, lack of a distinctive odor or taste, and the narrow close gills are the distinctive characters. This species should be added to the list of aspen-associated agarics for the Rocky Mountain area.

It is to be assumed from Kauffman's notes that a veil was present in the young fruiting bodies, but he did not state this as a fact. *H. salmonense* differs in having a thinly fibrillose stipe and a distinctly raphanoid odor and taste. Dr. B. B. Kanouse, who discovered this species, was for many years curator in the University of Michigan Herbarium.

38. Hebeloma salmonense sp. nov.

Illus. Figs. 39–40.

Pileus 3–6 cm latus, obtusus demum convexus vel plano-umbonatus, ad marginam fibrillosus, demum glaber, viscidus, pallide luteoalbus; odor et gustus raphanoinus. Lamellae albidae demum fulvae, confertae, angustae demum latae. Stipes 6–9 cm longus, 6–10 mm crassus, albus, tarde deorsum argillaceus, sparse fibrillosus. Sporae 9–13 × 6.5–8 μm, ± dextrinoideae, limoniformes. Basidia tetraspora. Cheilocystidia 36–58 × 7–10 × 6–7 × 7–9 μm, fusoide ventricosa, elongato-clavate vel cylindrico-subcapitata.

Specimen typicum in Herb. Univ. Mich. conservatum est, Smith 70148; legit prope Burgdorf, Idaho, 4 Sep 1964.

Pileus 3–6 cm broad, obtuse expanding to convex or plane to plano-umbonate, margin incurved at first and slightly fibrillose from a thin veil, soon glabrous, viscid, pale pinkish buff overall and only slightly darker as dried (whitish at first); odor and taste raphanoid.

Lamellae narrow becoming broad, close, adnate to adnexed, edges even, not beaded or spotted.

Stipe 6–9 cm long, 6–10 mm thick, equal, soon hollow, white, gradually becoming clay color from the base upward, thinly fibrillose from the remains of a white fibrillose veil, as dried ± concolorous with pileus.

Spores 9–13 × 6.5–8 μm, surface faintly marbled, pale clay color in KOH, ± dextrinoid; shape in profile inequilateral, in face view ovate to broadly ovate with the apex blunt.

Hymenium.—Basidia 4-spored, 7–9 (10) μm broad, projecting when sporulating. Pleurocystidia none. Cheilocystidia scattered 36–58 × 7–10 × 6–7 × 7–9 μm, cylindric-subcapitate, fusoid-ventricose with a ± enlarged apex or ± elongate-clavate and elongating considerably, surface ± viscid (as judged by adhering spores).

Lamellar and pilear tissues.—Lamellar trama typical for the genus. Cuticle of pileus an ixocutis to an ixolattice, the hyphae 1.5–3 μm diam, hyaline refractive in KOH, clamps present. Hypodermium hyaline revived in KOH, yellowish to reddish tawny in Melzer's, ± cellular, the cells 20–40 μm diam, some hyphal fragments also present, hyphal walls smooth and thin. Tramal hyphae ± typical for the genus (radial-interwoven and the hyphal cells ± inflated), merely yellowish in Melzer's.

Habit, habitat, and distribution.—Gregarious in brushy areas along an old ditch, French Creek Grade, Salmon River, Idaho, near the town of Burgdorf, September 4, 1964 (type, MICH).

Observations.—The raphanoid odor and taste and a slowly discoloring stipe are distinguishing field features. The odor and taste of *H. barrowsii* distinguishes that species from *H. salmonense* as do the more highly ornamented spores.

39. Hebeloma barrowsii sp. nov.

Pileus 2–4.5 cm latus, demum late convexus, glaber, viscidus, albus vel subalbidus; contextus subalbidus; sapor amarellus ("unpleasant"—Barrows); (odor "woody"—Barrows). Lamellae albae demum argillaceae, confertae, angustae. Stipes 3–5 cm longus, 4–6 mm crassus, albus, deorsum demum leviter umbrinus. Velum fibrillosum. Sporae 10–12.5 × 5–7 μm, sublimoniformes, tarde et leviter dextrinoideae. Basidia tetraspora. Cheilocystidia (34) 46–68 × 4–6 × 8–14 μm, elongato-clavata vel filamentosa vel capitata.

Specimen typicum in Herb. Univ. Mich. conservatum est, Barrows 3056; legit prope Hyde Park, elev. 8,000 ft, Santa Fe, New Mexico, Sep 1965.

Pileus 2–4.5 cm broad, ± convex becoming broadly convex, the margin at first incurved, glabrous, viscid, white, becoming dingy but as dried whitish ("Pale Pinkish Buff"). Context whitish, taste unpleasant, odor "woody" (Barrows).

Lamellae white becoming ± "Sayal Brown" and as dried cinnamon-tan, close, narrow, adnate, seceding, not beaded, edges even.

Stipe 3–5 cm long, 4–6 mm thick, equal, white (pallid overall when well dried), some slightly darker at the base, near apex furfuraceous, appressed fibrillose below. Veil very thin and all traces soon vanishing.

Spores 10–12.5 × 5–7 μm, minutely warty-roughened under high dry objective, ochraceous-pallid in KOH, slowly weakly dextrinoid; in profile view narrowly inequilateral and apex ± snoutlike in many, ovate to boat-shaped in face view.

Hymenium.—Basidia 4-spored, 8–10 μm broad, subcylindric and projecting prominently when sporulating. Pleurocystidia none. Cheilocystidia (34) 46–68 × 4–6 × 8–14 μm, elongate-clavate to filamentous-capitate, head ± oval in many; hyaline, thin-walled.

Lamellar and pilear tissues.—Lamellar trama typical of the genus. Cuticle of pileus a thin ixocutis of hyphae 1.5–2.5 μm diam, hyaline, refractive, smooth, clamped. Hypodermium hyaline in KOH, hyphoid (± distinct from the tramal hyphae by the width of the cells), in Melzer's yellowish (about like the tramal hyphae), some dextrinoid debris observed but this disappearing in a few minutes. Tramal hyphae typical of the genus. Clamps present.

Habit, habitat, and distribution.—Collected at Hyde Park, Santa Fe, New Mexico, in September of 1965 by Charles Barrows, no. 3056 (type, MICH).

Observations.—The spores are distinctly roughened under a high-dry objective, and the hypodermium shows some cell differentiation but no distinctive color in either KOH or Melzer's. There are indications of a veil on the stipe of well-dried specimens, which leads us to place the species in the subgenus *Hebeloma;* however, the cheilocystidia are typical of subgenus *Denudata. H. candidipes* has a number of characters in common with Barrows's species, but the features of the hypodermium readily distinguish them.

Section HEBELOMA

Subsection PRAEOLIDAE

Odor grata.

Typus: *Hebeloma praeolidum*

Odor fragrant.

KEY TO STIRPES

1. Spores 7–12 μm long .... Stirps *Praeolidum* (p. 89)
1. Spores 10–15 μm long .... Stirps *Fragrans* (p. 90)

Stirps PRAEOLIDUM

Spores 7–12 μm long.

KEY TO SPECIES

1. Growing under ash and soft maple on swampy ground
.... see *H. naucorioides* (i) (p. 171)
1. Growing under or near conifers 2
   2. Veil grayish; taste mild .... see 10. *H. brunneodiscum* (p. 39)
   2. Veil pallid to buff 3
3. Lamellae narrow; cheilocystidia 27–41 × 5–11 μm, cylindric to fusoid
   ventricose .... see 50. *H. pinetorum* (p. 102)
3. Lamellae broad at maturity 4
   4. Taste very disagreeable; odor heavy and sickening-sweetish in time
   .... 40. *H. praeolidum*
   4. Taste mild but odor strong and peculiar; spores ± 8–10 μm long
   .... see 21. *H. brunneomaculatum* (p. 52)

40. Hebeloma praeolidum sp. nov.

Illus. Pl. 4.

Pileus (1.5) 2–4 (6) cm latus, obtuse conicus demum late convexus vel umbonatus, viscidus, ad marginem fibrillosus, glabrescens; ad centrum pallide argillaceus vel luteobrunneus, ad marginem pallide griseobrunneus; odor graveolens; sapor valde subamarus (distinctissimus). Lamellae confertae, latae, luteobrunneae. Stipes 4–7 cm longus, (4) 5–7 mm crassus, deorsum brunneus, sursum pallide argillaceus. Sporae 9–12 × 5–6.5 μm, in "KOH" subleves, argillaceae. Basidia tetraspora. Cheilocystidia 36–50 × 5–8 μm, filamentosa vel ad basin ± ventricosa.

Specimen typicum in Herb. Univ. Mich. conservatum est, Smith 17169; legit prope Olympic Hot Springs, Olympic National Park, Washington, 22 Sep 1941.

Pileus (1.5) 2–4 (6) cm broad, obtusely conic with an inrolled margin, broadly convex to expanded umbonate at maturity, surface viscid and glabrous except for scattered fibrillose patches from the remains of the thin pallid buff cortina along the margin, on the disc "Cinnamon-Buff" or browner, the marginal area paler and grayer. Context thick in the disc (3–4 mm), tapered evenly to the margin,

"Pale Pinkish Buff" taste very disagreeable, odor heavy and sweetly aromatic (sickening in time).

Lamellae moderately broad (± 3 mm deep), moderately close (about 29 reach the stipe, lamellulae in 2 tiers, nearly free to (in age) ± sinuate, "Tawny-Olive" when covered by spores, edges even and not beaded or spotted.

Stipe 4–7 cm long, (4) 5–7 mm thick equal or enlarged downward, hollow, cortex pallid above and brownish near base, surface glabrous above, with scattered loose fibrils lower down (from the thin veil), "Pale Pinkish Buff" at the apex (yellowish pallid), darkening to cinnamon-brown from below upward in age.

Spores 9–12 × 5–6.5 μm, shape in profile view inequilateral, in face view ovate, faintly marbled under a high-dry objective (mounted in KOH) and ± clay color in groups; not dextrinoid.

Hymenium.—Basidia 4-spored. Pleurocystidia none. Cheilocystidia 36–50 (70) × 5–8 (11) μm, filamentous to slightly inflated in the lower (basal) area, or fusoid-ventricose with a long neck and obtuse apex (elongated fusoid-ventricose); some clavate cells 9–12 μm broad also present, all hyaline smooth and thin-walled.

Lamellar and pilear tissues.—Lamellar trama typical for the genus. Cuticle of pileus an ixolattice (possibly originating as an ixotrichodermium), the hyphae 2–4.5 μm diam, the walls refractive, clamps present. Hypodermium hyphoid, dingy ochraceous in mounts of fresh material, dull yellow-brown as revived in KOH. Tramal hyphae of pileus typical for the genus.

Habit, habitat, and distribution.—Scattered on moss, Olympic Hot Springs, Olympic National Park, Washington, September 22, 1941, Smith 17169 (type, MICH). It is also known from northern Idaho.

Observations.—This species is very distinct by reason of the pronounced odor and taste, thin veil, medium-sized spores, essentially fusoid-ventricose cheilocystidia and alutaceous pilei. *Hebeloma sacchariolens* sensu Moser (1978) has distinctly larger spores. No tendency for the stipe to radicate was noted in the North American collections, and a veil is clearly evident. Bruchet (1970) placed *H. sacchariolens* in the *Denudata* and commented on its cheilocystidia.

Stirps FRAGRANS

Spores 10–15 μm long.

KEY TO SPECIES

1. Spores dextrinoid 2
1. Spores not dextrinoid 3
    2. Taste bitter; veil pale buff .... 41a. *H. fragrans* var. *intermedium*

41. Hebeloma fragrans sp. nov. var. fragrans

Pileus 1.5–3 cm latus, obtusus demum convexus vel planus, glaber, vel ad marginem leviter fibrillosus, subviscidus, ad centrum fuscobrunneus, tarde pallidior. Context fragilis, fragrans, sapor mitis. Lamellae subdistantes, latae demum ventricosae, cinnamomeo-argillaceae. Stipes 2–3.5 cm longus, 3–4.5 mm crassus, fragilis, cinereobrunneus, brunnescens; velum cinereo-argillaceum. Sporae 10–15 × 6–7 μm, non dextrinoideae; inequilaterales, Cheilocystidia 42–73 × 4–6 μm, ± cylindrica vel filamentosa, vel 38–67 × 6–12 × 4–6 μm, fusoide ventricosa, hyalina.

Specimen typicum in Herb. Univ. Mich. conservatum est, Smith 90594; legit prope Elk Camp, Burnt Mt., Pitkin County, Colorado, 28 Aug 1980, Evenson, Mitchel, and Smith.

Pileus 1.5–3 cm broad, obtuse to convex, becoming plane, surface soon glabrous or margin retaining faint patches of grayish buff veil remnants, subviscid, disc fuscous-brown, margin dull tawny to paler, opaque at all times, when faded pale reddish tan except for fuscous-brown disc. Context thin and fragile, odor fragrant, taste slight; $FeSO_4$ blackish in base of stipe.

Lamellae subdistant, depressed-adnate, broad and in age ventricose, ± "Sayal Brown" at maturity.

Stipe 2–3.5 cm long, 3–4.5 mm thick, equal, fragile, pale grayish brown overall and fibrillose from grayish buff veil, darkening from the base upward, no veil zones evident.

Spores 10–15 × 6–7 μm, appearing smooth in KOH and punctate in Melzer's; not dextrinoid when fresh or as dried, inequilateral in profile view, ovate in face view.

Hymenium.—Basidia 4-spored, 8–10.5 μm broad near apex. Pleurocystidia none. Cheilocystidia 42–73 × 4–6 μm and cylindric to filamentous, and some 38–67 × 6–12 × 4–6 μm, fusoid-ventricose and apex not enlarged, hyaline in KOH, scattered along the edge as well as in groups.

Lamellar and pilear tissues.—Lamellar trama typical of the genus. Cuticle of pileus a thin ixocutis of hyphae 1.5–3 (4) μm diam, branched sparingly, clamps present. Hypoderm hyphoid to cellular, walls not highly colored in KOH or in Melzer's. Pileal trama typical of the genus (no red tints developing in Melzer's medium).

Habit, habitat, and distribution.—Gregarious under spruce and fir, Elk Camp, Burnt Mt., Pitkin County, Colorado, August 28, 1980 (type, MICH 90594). (Smith 90620, 90629, 90630 are additional collections.)

Observations.—The stipe was not white at first in any of our collections. The grayish buff veil, lack of a bitter or raphanoid taste, and the presence of a fragrant odor (which fades soon after the collection is unwrapped) distinguish this species. In one collection, Smith 90630, the taste was slightly bitter.

41a. Hebeloma fragrans var. intermedium var. nov.

Pileus 2–4.5 cm latus, convexus demum planus, subviscidus, rufobrunneus demum sordide fulvus; odor fragrans, gustus amarus. Lamellae latae demum ventricosae, confertae, subfulvae. Stipes 3–6 cm longus, ± 4 mm crassus, brunnescens; velum pallide subochraceum. Sporae 11–14 × 6–7 μm, dextrinoideae. Cheilocystidia versiforme: (1) fusoid-ventricosa et 43–70 × 7–12 × 3.5–5 μm; (2) filamentosa et 28–70 × 3–6 μm; (3) rare subsaccata.

Specimen typicum in Herb. Univ. Mich. conservatum est, Smith 90636; legit prope Burnt Mt., Pitkin County, Colorado, 30 Aug 1980.

Pileus 2–4.5 cm broad, convex becoming plane, subviscid, soon glabrous, dull reddish brown fading to pale tawny, opaque when moist. Context watery brownish fading to pallid; odor distinctly fragrant, taste bitter.

Lamellae broad, adnate, close, bright "Sayal Brown" (cocoabrown) edges eroded, not spotted and not beaded.

Stipe 3–6 cm long, ± 4 mm thick, equal, brunnescent from the base upward, at first thinly fibrillose from the pale buff veil (lacking zones or patches of veil fibrils).

Spores 11–14 × 6–7 μm, in profile inequilateral, in face view ovate, in KOH pale brownish ochraceous, in Melzer's dextrinoid and appearing smooth.

Hymenium.—Basidia 4-spored, yellow in Melzer's (on fresh material as well as dried). Pleurocystidia none. Cheilocystidia versiform: (1) fusoid ventricose and 43–70 × 7–12 × 3.5–5 μm; (2) filamentous and 28–70 × 3–6 μm; (3) rarely subsaccate to broadly fusoid ventricose, all hyaline and thin-walled.

Lamellar and pilear tissues.—Lamellar trama typical of the genus (yellowish in Melzer's). Cuticle of pileus a well-defined ixolattice of hyphae 2–4 μm diam, hyaline in KOH and clamped. Hypoderm cellular (in radial sections), hyphal walls yellow in Melzer's (on fresh material). Tramal hyphae yellowish in Melzer's, radial-interwoven and typical for the genus.

Habit, habitat, and distribution.—Solitary to scattered under spruce, Burnt Mt. area, Pitkin County, Colorado, August 30, 1980, Vera Evenson and A. H. Smith 90636 (type, MICH).

Observations.—The cellular hypoderm, bitter taste, and spores 11–14 × 6–7 μm, distinguish var. *intermedium* from the European *H. sacchariolens.* Since a bitter taste among nonraphanoid species appears to be a valid taxonomic character in subgenus *Hebeloma,* and since evidence of intergradation with var. *fragrans* has been found (see Smith 90630), we place Smith 90636 and others, as a variety of *H. fragrans.*

42. Hebeloma boulderense sp. nov.

Illus. Pl. 12.

Pileus 2–3 cm latus, convexus demum planus, ad marginem sparse fibrillosus, glabrescens, subviscidus, cinnamomeus. Contextus albus tactu pallide ochraceus; sapor mitis; odor suaveolens. Lamellae confertae, adnatae, latae, avellaneae demum cinnamomeae. Stipes 3–4 cm longus, 3–4 mm crassus, deorsum lutescens, sursum albopruinosus. Velum sparsim, albidum. Sporae 10–13.5 × 6–7.5 μm, leviter dextrinoideae, subfusiformes. Cheilocystidia 48–70 × 5–8 × 3–4 (5) μm, fusoide ventricosa.

Specimen typicum in Herb. Denver Bot. Gardens est, DBG-7959; legit prope Hesse, Boulder County, Colorado, 20 Aug 1979, Vera Evenson.

Pileus 2–3 cm broad, broadly campanulate to convex, margin sharp and even, extending over the edge as a sterile membrane; surface glabrous except near the margin where whitish veil remnants are seen as scattered patches; slightly viscid; disc "Sayal Brown" changing to pale ochraceous to the margin. Context white, thin, staining slightly ochraceous; taste not distinctive; odor of flowers (heavy and fragrant), $FeSO_4$ on base of stipe greenish.

Lamellae close, moderately broad, adnate, beaded in young specimens but not spotted even in age, avellaneous at first and becoming pale "Sayal Brown."

Stipe 3–4 cm long, 3–4 mm thick near apex, equal, white pruinose at apex, becoming dusted with spores in midsection, gradually darkening to ochraceous toward the base.

Spores 10–13.5 × 6–7.5 μm, rugose, ornamentation obvious

under oil-immersion lens, inequilateral in profile varying to subfusoid, in face view ovate, slightly dextrinoid.

Hymenium.—Basidia typical for the genus, 4-spored. Pleurocystidia absent. Cheilocystidia 48–70 × 5–8 × 3–4 (5) μm, elongate-fusoid ventricose, neck flexuous, apex blunt.

Lamellar and pilear tissues.—Lamellar trama parallel, hyaline in KOH, yellowish in Melzer's. Cuticle of pileus an ixolattice of hyphae 2–3 μm diam, often branched, clamp connections present. Hypodermium cellular, scarcely distinctive in KOH (slightly ochraceous). Pilear trama very pale ochraceous in KOH, interwoven, most hyphae enlarged to 10–12 μm diam, not reddish tinted in Melzer's.

Habit, habitat, and distribution.—Scattered under spruce and fir, west of Hesse town site, Boulder County, Colorado, August 20, 1979, Vera Evenson (type, DBG).

Observations.—This species is close to *H. fragrans* (which see), but its stipe is greenish at the base in $FeSO_4$, the context stained slightly ochraceous, the hypodermium is merely yellowish as revived in KOH, the veil is white, and the spore ornamentation is readily visible under an oil-immersion objective.

43. Hebeloma subsacchariolens sp. provisiorum

Pileus ± 4.5 cm broad, broadly convex, surface dry and under a lens thinly matted-fibrillose, or near the margin streaked with aggregations of fibrils, the edge decorated with patches of pale buff veil fibrils; disc ± pale dull brown, margin at first grayish brown. Context thin, white; odor fragrant, taste bitter but soon fading. KOH on cuticle no reaction: $FeSO_4$ staining the base of the stipe green instantly.

Lamellae broad, subdistant, adnate, becoming ventricose, pale dull tan; edges eroded but neither beaded nor spotted.

Stipe ± 6 cm long and ± 1 cm thick, soon hollow; surface dark tan below, paler near apex, at first with a thin coating of pale buff fibrils from the veil but no annular zone or zones evident.

Spores 10–14 × 6.5–8.5 (10) μm, smooth, clay color in KOH; shape in profile view inequilateral and many with a snout, in face view broadly elliptic to subelliptic or subglobose; pale reddish in Melzer's (weakly dextrinoid); often with a large central globule (as mounted in KOH).

Hymenium.—Basidia 4-spored, 32–40 × 9–14 μm clavate, containing globules, in mass reddish orange in Melzer's. Pleurocystidia none. Cheilocystidia variable: basically with a ventricose area midway toward base or near the base, neck usually finally greatly elongated and either tubular or the walls flexuous; apex obtuse to subcapitate; yellowish to pale clay color or merely hyaline as revived in KOH;

elongate cells (filamentous or cylindric) 38–70 × 4–6 μm, the fusoid-ventricose cells 31–60 × 6–12 × 4–6 μm.

Lamellar and pilear tissues.—Cuticle of pileus an ixocutis of appressed hyaline 2.5 μm wide and hyaline in KOH, clamped. Hypodermium of inflated hyphal cells (up to 30 μm more or less but the tissue essentially hyphoid, rusty brown and with distinct incrustations as revived in KOH. Tramal hyphae reddish in Melzer's but soon fading to ochraceous.

Habit, habitat, and distribution.—Solitary under spruce and fir, Elk Camp, Burnt Mt., Pitkin County, Colorado, September 3, 1977, Smith 90215.

Observations.—The above collection at a glance appears to be a basidiocarp of *Hebeloma sacchariolens* Quélet, but has a well-developed veil. Quélet's species has been consistently placed in subgenus *Denudata* (see Bruchet 1970 and also Moser 1978). The distinguishing features of *H. subsacchariolens* are the veil, the large (especially the wide) spores, the fragrant odor, the ventricose cheilocystidia, and the peculiar taste. It appears to belong in the vicinity of *H. fragrans* and *H. perplexum* but the wide spores appear to distinguish it from both. We include it here as a "provisional" species.

44. Hebeloma perplexum sp. nov.

Pileus 2.5–3.5 cm latus, obtusus demum convexus vel planus, subviscidus, ad marginem fibrillosus, demum glaber, spadiceus. Contextus fragrans; sapor perplexum. Lamellae confertae, latae, adnatae, brunneolae demum sordide cinnamomeae. Stipes 4–5 cm longus, 4–6 mm crassus, aequalis vel ad basin incrassatus, deorsum valde fulvescens. Velum albidum demum griseum. Sporae 10–13 × 6–7 μm, inequilaterales, leves, non dextrinoideae. Cheilocystidia (40) 53–75 × 6–9 × 4–6 (8) μm, elongate fusoid-ventricosa, vel (rare) saccata, 9–12 μm lata.

Specimen typicum in Herb. Univ. Mich. conservatum est, Smith 90498; legit prope Elk Camp, Burnt Mt., Pitkin County, Colorado, 7 Aug 1980, Evenson and Smith.

Pileus 2.5–3.5 cm broad, obtuse to convex in age, broadly convex to ± plane, surface subviscid and soon dry, margin at first with thin patches of veil fibrils, color of cap "Snuff Brown" to dull grayish brown on margin, disc "Bister" or paler (± date brown). Context with a fragrant odor; taste disagreeable and peculiar; FeSO₄ quickly black on base of the stipe.

Lamellae close, broad, adnate, pale dull cinnamon when mature, neither spotted nor beaded.

Stipe 4–5 cm long, 4–6 mm thick, equal to the slightly enlarged base, soon rusty brown from the base upward, grayish fibrillose in age. Veil white at first, slowly grayish pallid; apex ± faintly white-scurfy.

Spores 10–13 × 6–7 μm, in profile inequilateral, ovate in face view, weakly ornamented (appearing smooth), not dextrinoid.

Hymenium.—Basidia 4-spored, the layer reddish in Melzer's and fading to orange. Pleurocystidia none or only near the gill edge and similar to the cheilocystidia. Cheilocystidia (40) 53–75 × 6–9 × 4–6 (8) μm, fusoid-ventricose, some spathulate at apex, some saccate and 9–12 μm or more wide.

Lamellar and pilear tissues.—Lamellar trama typical for the genus to ± orange in Melzer's. Cuticle of pileus an ixolattice, hyphae tubular and 2–4 μm diam, smooth, clamped. Hypodermium hyphoid, rather indistinct, brownish in KOH as revived, redder in Melzer's, incrustations not conspicuous. Tramal hyphae of pileus radial-interwoven but with greatly inflated cells scattered through the tissue, the hyphae orange-ochraceous in Melzer's or at first more reddish orange.

Habit, habitat, and distribution.—Under spruce, Burnt Mt., Pitkin County, Colorado, August 7, 1980, Evenson and Smith (type, MICH).

Observations.—The fragrant odor, peculiar and disagreeable taste, whitish then grayish veil, medium large spores and the strongly fulvescent stipe are a distinctive combination of features.

Section HEBELOMA

Subsection MESOSPORAE subsect. nov.

Sporae 7–10 (11) × 4–7 μm.

Typus: *Hebeloma velatum*

Spores 7–10 (11) × 4–7 μm. (If spores measure 9–12 μm long see subsect. *Magnisporae* also.)

KEY TO SPECIES

1. Stipe not darkening in lower part by maturity 2
1. Stipe darkening (often slowly) from the base upward 4
    2. Taste bitter-farinaceous; pileus orange-brown when moist
    .... 45. *H. aurantiellum*
    2. Not as above 3
3. Cheilocystidia 26–33 × 8–12 μm .... 46. *H. immutabile*
3. Cheilocystidia 40–67 × 4.5–7 μm .... see 78. *H. fastibile* (p. 139)

45. Hebeloma aurantiellum sp. nov.

Illus. Fig. 38.

Pileus 3–3.5 cm latus, late convexus vel umbonatus, aurantio-cinnamomeus demum cinnamomeus, in siccitate vinaceobrunneus; gustus farinaceo-amarus. Lamellae confertae, latae, adnexae. Stipes 9–10 cm longus, 3–4 mm crassus, sericeo-fibrillosus, pallidus. Velum pallide alutaceum. Sporae 7–9 × 5–5.5 μm, dextrinoideae, in "KOH" pallide luteae, inequilaterales, subleves. Basidia tetraspora. Cheilocystidia 12–22 × 3–6 × 1.5–2.5 × 3–4 μm, ± tibiiformes. Cuticula pileorum ixotrichoderma est.

Specimen typicum in Herb. Univ. Mich. conservatum est, Smith 55477; legit prope Grants Pass, Oregon, 11 Nov 1956.

Pileus 3–3.5 cm broad, convex-umbonate, glabrous, slimy viscid, when moist "Mikado Brown" to "Cinnamon" but duller than both, more or less "Army Brown" (vinaceous-brown) when dried (in the herbarium), opaque at all times. Context thin, paler than the cuticle, odor slight, taste bitter-farinaceous.

Lamellae broad, close, adnexed and seceding, near "Pinkish Cinnamon" as dried, not beaded and not spotted.

Stipe 9–10 cm long, 3–4 mm thick, equal, typically pallid and silky fibrillose as well as having patches of buff-colored veil fibrils (one specimen had a brownish but water-soaked stipe).

Spores 7–9 × 5–5.5 μm, in KOH pale ochraceous, dextrinoid,

surface minutely marbled, shape in profile view inequilateral, in face view ovate, apex blunt.

Hymenium.—Basidia 4-spored, 6–7 μm wide near apex. Pleurocystidia none. Cheilocystidia very small, 12–22 × 3–6 × 1.5–2.5 × 3–4 μm, ventricose near base, neck narrow and with a slight apical capitellum, hyaline.

Lamellar and pilear tissues.—Lamellar trama typical for the genus. Cuticle of pileus a well-developed ixotrichodermium collapsing to an ixolattice, the hyphae about 1 μm in diam, clamped, hyaline, refractive, often branched. Hypodermium hyphoid, delimited as a ± clay-colored zone (in KOH mounts), redder in Melzer's and dextrinoid debris evident. Tramal hyphae typical of the genus.

Habit, habitat, and distribution.—In a mixed conifer–hardwood forest, Grants Pass, Oregon, November 11, 1956 (type, MICH).

Observations.—The important features of this species are the bitter-farinaceous taste, the dull orange-brown pileus when fresh, the small ± tibiiform cheilocystidia, small spores and the ixotrichodermium forming the cuticle of the pileus. The cheilocystidia and spores readily distinguish it from *H. pascuense* Pk. but the pigmentation of the two is quite similar, as is evidenced by Peck's plate (1900). *H. aurantiellum* has the small spores of *H. pumilum* Lange but has a buff-colored veil along with other differences.

46. Hebeloma immutabile sp. nov.

Pileus 2–4 cm latus, convexus vel planus, glutinosus, ad marginem minute squamulosus vel appendiculatus, glabrescens, obscure vinaceobrunneus; odor et gustus mitis. Lamellae confertae, latae, adnatae, subcinnamomeae. Stipes 4–7 cm longus, 2.5–4 mm crassus, sericeofibrillosus, brunneolus sed deorsum non brunnescens). Velum sparsim, evanescens. Sporae 8–10 × 5–5.5 μm, in "KOH" fulvae, valde dextrinoideae, verruculosae, inequilaterales. Cheilocystidia 26–38 × 8–12 μm, utriformia vel clavata. Cuticula pileorum ixotrichoderma est. Hypoderma ± cellulosa.

Specimen typicum in Herb. Univ. Mich. conservatum est, Smith 17538; legit prope Port Angeles, Washington, 4 Oct 1941.

Pileus 2–4 cm broad, convex becoming plane, margin at first incurved, surface glabrous to the marginal area where at first floccules from the veil occur, the margin at first slightly appendiculate, surface slimy-viscid, color ± "Army Brown" to "Pecan Brown" and drying a dingy vinaceous-brown over the margin and the disc reddish brown. Odor and taste mild.

Lamellae broad, close, adnate, as dried a dull rusty cinnamon; not spotted and not beaded when fresh, edges even.

Stipe 4–7 cm long, 2.5–4 mm thick, equal, fibrillose-silky to apex, no veil-line evident and no appreciable zones or patches of veil material present, brownish and concolorous throughout but not darkening from the base upward.

Spores 8–10 × 5–5.5 μm, rusty cinnamon in groups in KOH, paler singly, (± ochraceous), strongly dextrinoid, rugulose (± as in *Cortinarius*), shape in profile view distinctly inequilateral, ovate in face view, blunt at apex.

Hymenium.—Basidia 4-spored, 7.5–9 μm broad near apex, clavate. Pleurocystidia none. Cheilocystidia 26–38 × 8–12 μm, ± utriform to ± clavate, apparently the surface gelatinous (judged by the manner in which spores adhere to the surface); often with an internal apical deposit.

Lamellar and pilear tissues.—Lamellar trama typical for the genus. Cuticle of pileus a typical ixotrichodermium, the hyphae 1.5–2 μm diam, sparingly branched, walls refractive, clamp connections present. Hypodermium hyphoid (but appearing cellular in places), walls of the elements bay-brown in Melzer's. Tramal hyphae typical of the genus but in thick sections the layer pale brown, as revived in Melzer's flushed reddish for a time.

Habit, habitat, and distribution.—Gregarious under conifers, Mt. Angeles, Olympic Mountains, Washington, October 4, 1941 (type, MICH).

Observations.—The strongly dextrinoid distinctly roughened spores, broad short cheilocystidia, well-developed ixotrichodermium, the generally vinaceous-brown pileus, and nondarkening stipe make this a very distinctive species.

47. Hebeloma maritinum sp. nov.

Illus. Figs. 36–37.

Pileus 3–4.5 cm latus, convexus vel subplanus, ad marginem fibrillosus, glabrescens, vinaceobrunneus vel ± avellaneus, odor et gustus mitis. Lamellae avellaneae demum vinaceobrunneae, latae, confertae, adnexae. Stipes 5–7 cm longus, 4–8 mm crassus, cinereus, fibrillosus, brunnescens. Sporae 7–9 × 4.5–5.5 μm, subleves, in "KOH" pallide ochraceae, non dextrinoideae. Cheilocystidia fusoid-ventricosa, 40–60 × 6–9 × 3–4 (5) μm.

Specimen typicum in Herb. Univ. Mich. conservatum est, Smith 9008; legit prope Crescent City, California, 22 Nov 1937.

Pileus 3–4.5 cm broad, convex becoming plane with the margin spreading and more or less wavy, margin inrolled at first and then minutely tomentose, marginal area decorated with evanescent patches of fibrils representing the remains of a thin fibrillose veil; color "Verona Brown" on the disc, margin "Avellaneous" and somewhat zoned, becoming paler in age. Context concolorous with pileus, thick, odor and taste not distinctive.

Lamellae "Avellaneous" at first (pinkish gray), soon "Army Brown" and with a glaucous tinge in age, broad, crowded, adnexed.

Stipe 5–7 cm long, 4–8 mm thick, equal, stuffed then hollow, pallid grayish and appressed fibrillose, apical region faintly pruinose, becoming bister from the base upward in age.

Spores 7–9 × 4.5–5.5 μm, in profile view inequilateral, ovate in face view, very minutely marbled, pallid in KOH, not dextrinoid (pale ochraceous in Melzer's).

Hymenium.—Basidia 4-spored, 23–26 × 5–6 μm. Pleurocystidia none. Cheilocystidia fusoid-ventricose with greatly elongated necks and the apex obtuse, 40–60 × 6–9 × 3–4 (5) μm, thin-walled, wall in neck often flexuous, hyaline in KOH.

Lamellar and pilear tissues.—Lamellar trama typical for the genus. Cuticle of pileus an ixocutis, the hyphae 2–3 μm diam, interwoven (possibly originating as an ixotrichodermium), walls refractive in KOH mounts. Hypodermium hyphoid, dingy brownish in KOH. Clamp connections present.

Habit, habitat, and distribution.—Gregarious under fir on moss, Crescent City, California, November 22, 1937 (type, MICH).

Observations.—The combination of small spores inequilateral in profile view, darkening stipe, thin pale gray veil, and hyphoid hypoderm together appear distinctive.

48. **Hebeloma amarellum** sp. nov.

Illus. Figs. 34–35.

Pileus 3–7.5 latus, late convexus, glaber, viscidus, rufocinnamomeus, ad marginem albofibrillosus. Velum sparsim. Lamellae confertae vel subdistantes, adnatae, demum obscure vinaceobrunneae. Stipes 3–6 cm longus, 4–9 mm crassus, sursum albidus, deorsum demum "Bister" (sordide spadiceus), sursum squamulosus, deorsum fibrillosus, glabrescens. Sporae 7–9 × 4.5–5 μm, dextrinoideae, subleves, in "KOH" argillaceae, inequilaterales. Cheilocystidia 18–30 × 3.5–6 μm, filamentosa vel fusoid-ventricosa. In pratis in stercore vaccino.

Specimen typicum in Herb. Univ. Mich. conservatum est, Smith 9339; legit prope Crescent City, California, 3 Dec 1937.

Pileus 3–7.5 cm broad, obtuse to convex, margin inrolled and faintly pruinose to cottony-fibrillose, expanding to broadly convex in age, surface glabrous, viscid, "Verona Brown" to "Cinnamon," in some the disc a darker reddish brown, not truly hygrophanous but gradually becoming pinkish buff, sometimes radially rugulose. Veil rudimentary, evident on buttons in which the cap margin was ± 1 mm from the stipe. Context whitish to pallid, thin, taste bitterish, odor faint and not distinctive.

Lamellae broad, close to subdistant, adnate, becoming adnexed, pallid brownish when young, dark vinaceous-brown ("Rood's Brown" to "Walnut Brown") when mature.

Stipe 3–6 cm long, 4–9 mm thick, equal, solid or with a narrow pith, whitish above, brownish below, becoming bister (spadiceous) from the base upward, apex innately squamulose at first, more fibrillose to fibrillose-squamulose downward, veil rudimentary, pallid.

Spores 7–9 × 4.5–5 μm, in profile view ± inequilateral, in face view ovate, pale clay color in KOH, dextrinoid, minutely roughened.

Hymenium.—Basidia 4-spored, 22–27 × 5–6 μm, projecting when sporulating. Pleurocystidia none. Cheilocystidia small, 18–30 × 3.5–6 μm, filamentous to ventricose at base and with a flexuous neck, apex obtuse, hyaline to yellowish in KOH, agglutinating.

Lamellar and pilear tissues.—Lamellar trama typical for the genus. Cuticle of pileus a narrow ixocutis of hyphae 2–4 μm diam, hyaline, walls refractive, clamps present, hyphae much branched. Hypodermium intermediate as to type, yellow in KOH.

Habit, habitat, and distribution.—Gregarious on cow dung in a pasture, Crescent City, California, December 3, 1937 (type, MICH).

Observations.—This is a very distinct species by virtue of its habitat alone, but if habitat is disregarded, the bitter taste, dark vinaceous-brown gills and the small dextrinoid spores amply distinguish it.

49. **Hebeloma olympianum** sp. nov.

Pileus 1–2.5 cm latus, obtusus deinde campanulatus vel plano-umbonatus, tenuiter fibrillosus, glabrescens, viscidus, cinnamomeus; odor et gustus mitis. Lamellae pallidae demum brunneolae deinde cinnamomeae, confertae, latae, adnatae. Stipes 2–4 cm longus, 2.5–3.5 mm crassus, tenuiter fibrillosus, glabrescens, deorsum brunnescens. Sporae 7–9 × 4–5.5 μm, minute verruculosae, leviter dextrinoideae, inequilaterales. Cheilocystidia (18) 22–27 × 4–7 μm, ± filamentosa vel anguste clavata. Cuticula pileorum ixotrichoderma est.

Specimen typicum in Herb. Univ. Mich. conservatum est, Smith 17972; legit prope Storm King Mt., Olympic National Park, Washington, 16 Oct 1941.

Pileus 1–2.5 cm broad, obtuse to convex and with margin incurved, expanding to campanulate to plano-umbonate, margin at first faintly fibrillose from veil fibrils, soon glabrous overall, viscid; color "Sayal Brown" (dull cinnamon) to the "Cinnamon-Buff" margin (slightly paler than the disc). Context thin, brownish, odor and taste mild.

Lamellae broad, close adnate, pallid then brownish and finally ± "Sayal Brown," edges even and neither beaded nor spotted.

Stipe 2–4 cm long, 2.5–3.5 mm thick, equal, faintly fibrillose from the thin veil but no annular zone (veil-line) present (just a few patches or a partial zone present), becoming dark brown over the lower portion, pallid and silky above.

Spores 7–9 × 4–5.5 μm, roughened, dingy pale buff in KOH, ± dextrinoid; shape in profile view distinctly inequilateral, in face view ovate.

Hymenium.—Basidia 4-spored, 6.5–8 μm broad. Pleurocystidia none. Cheilocystidia small, (18) 22–27 × 4–7 μm, filamentous to narrowly clavate (possibly elongating somewhat), hyaline, thin-walled and difficult to revive.

Lamellar and pilear tissues.—Lamellar trama typical for the genus. Cuticle of pileus an ixotrichodermium collapsing to an ixolattice, the hyphae ± 1.5 μm diam, refractive, sparsely branched, clamped. Hypodermium hyphoid, rusty brown in KOH, bay-red in Melzer's, walls not heavily encrusted. Tramal hyphae typical of the genus. Clamps present.

Habit, habitat, and distribution.—On soil along a trail, Storm King Mt., Olympic National Park, Washington, October 16, 1941 (type, MICH).

Observations.—The hyphoid hypodermium and the distinct ixotrichodermium, the short cheilocystidia and small dextrinoid spores are the important microscopic features. The field characters are somewhat routine: lack of an odor and taste, the stipe darkening at the base, and the pale cinnamon pileus.

50. Hebeloma pinetorum sp. nov.

Pileus 1–2.5 (3) cm latus, demum campanulatus vel plano-umbonatus, sparsim fibrillosus, glabrescens, glutinosus, pallide argillaceus; odor fragrans. Lamellae angustae, confertae, adnatae, luteobrunneae. Stipes 2–3.5 cm longus, 2.5–3 mm latus, tenuiter fibrillosus, glabrescens, deorsum brunnescens. Sporae 9–11 × 6–7 μm, subleves in "KOH," dextrinoideae, inequilaterales. Cheilocystidia 27–41 × 8–11 μm, subcylindrica vel fusoid-ventricosa. Cuticula pileorum ixotrichoderma est. Hypoderma "hyphoid."

Specimen typicum in Herb. Univ. Mich. conservatum est, Smith 24799; legit prope Twin Bridges, Mt. Hood, Oregon, 18 Oct 1946.

Pileus 1–2.5 (3) cm broad, obtuse to obtusely campanulate, with an incurved margin, becoming expanded-umbonate to plane, the margin at the youngest stages with the scattered remains of a fibrillose veil, soon glabrous, surface slimy-viscid, opaque, cinnamon-buff to pinkish buff and retaining these colors in drying. Context thin, odor fragrant (taste not recorded but assumed to be mild).

Lamellae narrow, crowded, adnate, edges even, not beaded, at first pallid but becoming pinkish buff and finally about "Buckthorn Brown."

Stipe 2–3.5 cm long, about 2.5–3 mm thick, equal, pallid and thinly floccose from a slight veil, becoming brown at the base and then slowly upward.

Spores 9–11 × 6–7 μm, very faintly marbled, clay color to snuff brown in KOH (paler individually), dextrinoid, in profile view broadly inequilateral and with a very blunt apex, shape in face view broadly ovate.

Hymenium.—Basidia 4-spored, 7–9 μm broad, projecting when sporulating. Pleurocystidia none. Cheilocystidia inconspicuous, 27–41 × 8–11 μm, subcylindric to ± fusoid-ventricose, ± clavate when young, hyaline, thin-walled, not agglutinated.

Lamellar and pilear tissues.—Lamellar trama typical for the genus. Cuticle of pileus an ixotrichodermium collapsing to an ixolattice, composed of hyphae 1–1.5 μm diam, hyaline, refractive, clamped, branching fairly frequent. Hypodermium scarcely differentiated (hyphoid), merely ± clay color in KOH. Tramal hyphae compactly interwoven, hyaline, many minute granules developing in KOH mounts. Clamps present.

Habit, habitat, and distribution.—Gregarious under salal (*Gaultheria shallon* Pursh.), lodgepole pine, etc. Twin Bridges, Mt. Hood, Oregon, October 18, 1946 (type, MICH).

Observations.—The medium-sized blunt-nosed spores, cinnamonbuff pileus, darkening stipe and fragrant odor along with the relatively inconspicuous cheilocystidia are distinctive.

51. Hebeloma limacinum sp. nov.

Pileus 1–2.5 cm latus, convexus vel plano-umbonatus, glaber, glutinosus, triste fulvus, ad marginem subargillaceus; odor ± pungens;

sapor mitis. Lamellae angustae, subdistantes, pallide avellaneae demum rufobrunneae. Stipes 2–3 cm longus, 3–4 mm crassus, deorsum fusco-brunneus, sursum pallidus, sericeus. Velum pallidum tenuiter fibrillo-sum. Sporae 9–11.5 × 6–6.5 μm, dextrinoideae, inequilaterales. Chei-locystidia 36–65 (74) × 5–12 × 4–6 μm, subcylindrica vel ad basin ± ventricosa. Cuticula pileorum ixolattice est. Hypodermium cellulosum.

Specimen typicum in Herb. Univ. Mich. conservatum est, Smith 89110; legit prope Savage Lakes, Pitkin County, Colorado, 23 Aug 1978.

Pileus 1–2.5 cm broad, convex to plano-convex or umbonate, surface glabrous, slimy-viscid when fresh, disc dark tawny, margin pinkish buff or paler, pale pinkish buff overall when faded; odor slightly pungent, taste mild; $FeSO_4$ olive-fuscous on lower part of stipe.

Lamellae narrow, subdistant, adnate, pallid, avellaneous becoming reddish brown, neither beaded nor spotted.

Stipe 2–3 cm long, 3–4 mm thick, equal, soon dark brown at the base, pallid and silky at apex; veil remnants present as a few indistinct zones on lower half of stipe.

Spores 9–11.5 × 6–6.5 μm, in KOH dull clay color, in Melzer's weakly dextrinoid to tawny-red, surface obscurely mottled (under a high-dry objective), shape in profile inequilateral, in face view ovate, the apex blunt.

Hymenium.—Basidia 4-spored, 7–10 μm wide near apex. Pleuro-cystidia none. Cheilocystidia 36–65 (74) × 5–12 × 4–6 μm, cylindric down to a ventricose base, the neck finally greatly elongated, second-ary septa in neck rare; none forked at apex.

Lamellar and pilear tissues.—Lamellar trama typical of the genus except for the hymenium and subhymenium, becoming reddish orange in Melzer's. Cuticle of pileus a thick ixolattice of hyphae 2–4 μm diam, embedded in a slime matrix, walls refractive, dextrinoid debris moder-ately abundant in the layer, clamp connections present. Hypodermium a cellular layer (in tangential sections), intermediate as seen in radial sections, with patches of encrusting pigmented material on the cell walls or as wall thickenings, the walls and incrustations rusty brown in KOH, "Mars Orange" in Melzer's. Tramal hyphae typical of the genus (not red or orange in Melzer's).

Habit, habitat, and distribution.—On muck in a mixed conifer forest (pine and spruce), Savage Lakes area, Pitkin County, Colorado, August 23, 1978 (type, MICH).

Observations.—The slimy pileus when fresh, the narrow gills, the medium-sized spores, and lack of a distinctive odor and taste charac-terize this species.

52. Hebeloma subhepaticum sp. nov.

Illus. Figs. 28–29.

Pileus 2–4 (6) cm latus, convexus demum ± planus, glutinosus, incarnato-griseus demum obscure vinaceobrunneus vel subhepaticus, glaber; odor mitis, gustus ± farinaceus. Lamellae pallide incarnato-griseae demum rufobrunneae, latae, confertae, adnexae. Stipes (3) 5–6 cm longus, 3–8 (10) mm crassus, tenuiter fibrillosus, glabrescens, deorsum brunnescens, sursum pallidus. Sporae 8–10 × 5–5.5 μm, verruculosae, dextrinoideae, in "KOH" obscure fulveae, inequilaterales. Cheilocystidia 18–26 × 3–4 μm, subventricose vel tibiiformes. Cuticula pileorum ixotrichoderma est.

Specimen typicum in Herb. Univ. Mich. conservatum est, Smith 18029; legit prope Cape Flattery, Washington, 19 Oct 1941.

Pileus 2–4 (6) cm broad, convex with an incurved margin, becoming plane or nearly so; surface slimy-viscid, dingy vinaceous-buff at first, darkening through "Verona Brown" to "Warm Sepia" or liverbrown, a redder liver-brown when dried, margin long remaining paler. Context watery but firm, colored about like the pileus; odor and taste mild or taste ± farinaceous.

Lamellae broad, close, soon adnexed, pale vinaceous-buff becoming darker (like the pileus), as dried distinctive dark reddish brown, edges even and concolorous with the faces.

Stipe (3) 5–6 cm long, 3–8 (10) mm thick, equal, pallid above, surface ± floccose from remains of a thin evanescent veil, ± glabrescent, striate at times, soon dark brown at the base, hollow at maturity.

Spores 8–10 × 5–5.5 μm, minutely roughened, dextrinoid and in KOH dull rusty brown, paler singly, in profile view inequilateral, in face view ovate.

Hymenium.—Basidia 26–31 × 7–9 μm, 4-spored. Pleurocystidia none. Cheilocystidia very small, 18–26 × 3–4 μm, narrowly ventricose to tibiiform.

Lamellar and pilear tissues.—Lamellar trama hyaline in KOH, in Melzer's the whole mount flushed red on standing a short time. Cuticle of pileus an ixotrichodermium collapsing to an ixolattice, the hyphae ± 1.5 μm diam, branched, hyaline in KOH, clamps present. Hypodermium hyphoid, merely separated from trama proper as a colored reddish zone in Melzer's, as revived in KOH rusty brown.

Habit, habitat, and distribution.—Scattered to gregarious under Sitka spruce, Pacific Northwest: Wahington to California. Collections: Smith 56753, 56757, 56838, 57020, 57047, and (type, MICH 18029).

Observations.—The veil is very soon evanescent. The liver-colored pileus, farinaceous taste, vinaceous-buff lamellae at first, the darkening stipe and the ± tibiiform cheilocystidia make it a very distinct species.

53. Hebeloma parcivelum sp. nov.

Illus. Figs. 30–31.

Pileus 3–7.5 cm latus, conicus demum plano-umbonatus, viscidus, ad marginem gossypinus, glabrescens, pallide argillasceus demum brunneomaculatus; odor et gustus mitis. Lamellae avellaneae deinde cinnamomeae, adnatae, latae, confertae. Stipes 6–8 cm longus, 8–12 mm crassus, deorsum brunnescens. Velum sparsim, pallidum. Sporae 7–10 × 5–6 μm, subleves, dextrinoideae, inequilaterales. Cheilocystidia 20–30 × 5–6.5 μm, filamentosa vel fusoid-ventricosa. Hypodermium pileorum nullum

Specimen typicum in Herb. Univ. Mich. conservatum est, Smith 23703; legit sub *Alni,* prope Waupanitia Summit, Mt. Hood, Oregon, ± 3,000 ft elevation, 24 Sep 1946.

Pileus 3–7.5 cm broad, obtusely conic with an inrolled margin, expanding to plano-umbonate and the margin at times wavy, surface glabrous, slightly viscid, margin cottony at first, "Pale Pinkish Buff" overall or disc darker and near "Cinnamon-Buff," in age developing darker sordid brown spots. Context thick, concolorous with surface, odor and taste none.

Lamellae broad, close, bluntly adnate, "Avellaneous" at first, in age with a "Cinnamon" cast but always dull, edges even and not beaded.

Stipe 6–8 cm long, 8–12 mm thick, equal, solid, pale pinkish buff near apex, becoming dark "Saccardo's Umber" from the base upward and finally dark bister, a very slight veil present at first, leaving the surface somewhat fibrillose in the midportion, apical region pruinose.

Spores 7–10 × 5–6 μm, in profile view broadly inequilateral, in face view broadly ovate, very minutely roughened (marbled), pale dingy ochraceous in KOH, reddish tawny in Melzer's (± dextrinoid).

Hymenium.—Basidia 4-spored, 20–23 × 6–6.5 μm. Pleurocystidia none. Cheilocystidia small (20–30 × 5–6.5 μm), ± filamentous, some slightly ventricose at or near the base, apparently becoming more elongated in age.

Lamellar and pilear tissues.—Lamellar trama typical of the genus. Cuticle of pileus an ixocutis, hyphae 2.5–4 μm diam, walls refractive, the hyphae interwoven. Hypodermium not differentiated in fresh material. Clamp connections present.

Habit, habitat, and distribution.—Gregarious at edge of an alder bog near Waupanitia Summit, Mt. Hood, Oregon, at approximately 3,000 ft elevation, September 24, 1946 (type, MICH).

Observations.—The pale pinkish buff pileus becoming spotted with sordid brown spots, the thick stipe darkening near the base, the very thin veil, dextrinoid spores, and small ± filamentous cheilocystidia distinguish this species.

54. Hebeloma trinidadense sp. nov.

Pileus 3–9 cm latus, late convexus demum leviter depressus, glutinosus, albofloccosus deinde glaber, vinaceobrunneus vel "Mikado Brown" (± aurantio-cinnamomeus); odor et gustus mitis. Lamellae latae, adnatae, secedentes, brunneolae deinde cinnamomeae. Stipes 4–8 cm longus, 5–10 mm crassus, albidus, deorsum brunnescens, fibrillosus. Velum pallidum, sparsim, evanescens. Sporae 8–10 × 5–6 μm, verruculosae, dextrinoideae, inequilaterales. Cheilocystidia 26–41 × 6–8 × 4–5 μm, fusoid-ventricosa. Cuticula pileorum "ixolattice" est.

Specimen typicum in Herb. Univ. Mich. conservatum est, Smith 57047; legit prope Trinidad, California, 24 Dec 1965.

Pileus 3–9 cm broad, broadly convex with a decurved margin, edge curled in at first and whitish fibrillose-floccose from a thin delicate veil, becoming plane to slightly depressed on disc, glabrous, slimy-viscid, "Verona Brown" and in age darker, as dried "Mikado Brown" to a bright "Sayal Brown." Context brownish when moist, odor and taste not distinctive.

Lamellae ± broad, close, adnate but seceding, equal, pale brown becoming rusty cinnamon, edges even and not beaded.

Stipe 4–8 cm long, 5–10 mm thick, with a ball of debris held to the base by copious mycelium, white overall at first, soon darkening to brown at base and slowly darkening upward, fibrillose, silky near apex, with a thin fibrillose coating downward. Veil pallid, thin and all traces soon vanishing.

Spore deposit "Sayal Brown" as dried on the pilei. Spores 8–10 × 5–6 μm, distinctly roughened (under a high-dry objective), pallid brownish in KOH singly, in groups rusty cinnamon; shape in face view ovate, in profile view inequilateral dextrinoid.

Hymenium.—Basidia 4-spored, 6–7 μm wide, 20–26 μm long. Pleurocystidia none. Cheilocystidia small, ± fusoid-ventricose, 26–41 × 6–8 × 4–5 μm, not agglutinated, hyaline, thin walled.

Lamellar and pilear tissues.—Lamellar trama typical for the genus. Cuticle of pileus a thick ixocutis (over 100 μm thick), the hyphae much-branched, hyaline, refractive, 1.5–2 μm wide and clamped. Hypodermium intermediate as to type and not highly colored in Melzer's, no incrustations of significance noted. Tramal hyphae typical for the genus. Clamps present.

Habit, habitat, and distribution.—On waste ground, conifers nearby, Trinidad, California, December 24, 1965, White and Smith (type, MICH).

Observations.—The important characters appear to be the orange-brown pileus when young, mild odor and taste, thick stipe, the thin pallid veil, dextrinoid spores distinctly rough under a high-dry objective (43 ×), and the thick pilear ixocutis.

Section HEBELOMA

Subsection MAGNISPORAE subsect. nov.

Sporae 10 μm vel ultra longae, in lateralibus amygdalaceae.

Typus: *Hebeloma kuehneri*

Spores 10 μm or more long, inequilateral in profile.

KEY TO STIRPES

1. Stipe not staining or discoloring in the lower portion by late maturity
   .... Stirps *Coniferarum* (p. 108)
1. Stipe soon darkening at base or lower portion, then upward 2
    2. Spores dextrinoid (medium to dark reddish brown in Melzer's)
       .... Stirps *Kuehneri* (p. 115)
    2. Spores not dextrinoid, but in a mount a few may be found which become
       pale to ± reddish brown in 30 minutes; (dried specimens are most
       reliable for this test) .... Stirps *Oregonense* (p. 138)

Stirps CONIFERARUM

Stipe not darkening appreciably from the base up.

KEY TO SPECIES

1. Cheilocystidia fusoid-ventricose, the apices subcapitate; odor pungent
   .... 55. *H. pungens*
1. Not as above 2
    2. Cheilocystidia (some of them) cylindric-subcapitate; pileus slimy viscid;
       pilear cuticle an ixotrichodermium .... 56. *H. kelloggense*
    2. Not as above 3

3. Spores 10–15 × 6–7 (8) μm .... 57. *H. coniferarum*
3. Spores 9–12 × 5–6 μm 4
    4. Cheilocystidia pedicellate-clavate, enlarged apex 6–12 μm broad
    .... 58. *H. kemptonae*
    4. Not as above 5
5. Lamellae close; stipe 2–4 mm diam; spores truly dextrinoid .. 59. *H. griseocanum*
5. Lamellae distant, stipe 8–12 mm thick, spores not dextrinoid but may be ± tan
    after standing for 15 minutes after mounting ... 60. *H. pseudofastibile* var. *distans*

55. Hebeloma pungens sp. nov.

Pileus 3–5 (6) cm latus, paene "Verona Brown" juvenis, vetus paene "Tawny-Olive," sericocanescens, glabrescens, viscidus. Contextus albus; odor asper mordax, gustus mitis. Lamellae adnexae, pallidobrunnaceae, denique paene "Sayal Brown," confertae, latae. Stipes 4–6 cm longus, 8–12 mm crassus, albus, apice furfuraceus, fibrilloso-scabrosus infra deinde squamas recurvatus gerens. Velum sparsim, evanescens. Sporae (8) 9–11 (12) × (4.5) 5–6 μm, subfusoideae, rugulosae vel leviter calyptrateae. Pleurocystidia desunt. Cheilocystidia 30–54 × 5–8 μm, anguste utriformia. Cuticula pileorum ixocutis est. Hypodermium cellulare.

Specimen typicum in Herb. Univ. Mich. conservatum est, Smith 24293; legit prope Frog Lake, Mt. Hood National Forest, Oregon, 8 Oct 1946.

Pileus 3–5 (6) cm broad, obtuse to convex, with an incurved (inrolled) margin, becoming broadly convex, near "Verona Brown" when young, near "Tawny-Olive" in age, buttons canescent to silky, soon glabrous or a thin silky coating of fibrils persistent along the margin, viscid when wet, margin often splitting. Context thick on disc, tapering abruptly to the margin, white, odor pungent but soon gone (raphanoid but sharper, almost as sharp as garlic), taste not distinctive.

Lamellae broad, close, adnexed, pale cinnamon when young, darker in age, finally near "Sayal Brown," edges whitish and finally eroded, three tiers of lamellulae.

Stipe 4–6 cm long, (5) 8–12 mm thick, white or whitish, apex finely furfuraceous, somewhat fibrillose-scabrous lower down, the cuticle cracking into recurved scales, equal or at first clavate, solid. Veil pallid, thin.

Spores (8) 9–12 × (4.5) 5–6 μm, inequilateral in profile, subfusiform to subovate in face view, wall 0.2–0.3 μm thick, rugulose, some ± calyptrate, ochraceous in KOH, slowly dextrinoid.

Hymenium.—Basidia 4-spored, 28–34 × 6–8 μm. Pleurocystidia none. Cheilocystidia 30–54 × 5–8 μm, ventricose toward the base and apex subcapitate (± narrowly utriform).

Lamellar and pilear tissues.—Lamellar trama typical of the genus. Cuticle of the pileus a thin ixocutis, the hyphae narrow and hyaline. Hypodermium a brown cellular zone, cells not always sharply outlined. Stipe cuticle a zone of dry, pallid or grayish, thick-walled hyphae. Clamps present.

Habit, habitat, and distribution.—Caespitose on soil under pine, Frog Lake, Mt. Hood National Forest, Oregon, October 8, 1946, Gruber and Smith (type, MICH).

Observations.—This species is distinguished by its ± scaly stipe which does not become brown, the relatively narrow spores more distinctly ornamented than in our concept of *H. fastibile*, and a ± cellular hypodermium.

56. Hebeloma kelloggense sp. nov.

Pileus 3–6 cm latus, convexus demum late-umbonatus, glaber, glutinosus, obscure cinnamomeus, in siccati vinaceobrunneus, odor pungens. Lamellae latae, confertae, adnatae brunneolae demum subfulvae. Stipes 6–9 cm longus, 5–10 mm crassus, albus, non brunnescens; velum sparsum, fibrillosum. Sporae 9–12 × 5–6.5 μm, dextrinoideae, subleves, inequilaterales. Basidia tetraspora. Cheilocystidia 27–57 × 8–9 × 7–11 μm, cylindric-subcapitata, capitellum 8–9 μm diam; vel subclavatta vel contorta, et 27–36 × 6–7 × 7–11 μm, non elongata.

Specimen typicum in Herb. Univ. Mich. conservatum est, Smith 73592; legit prope Kellogg, Idaho, 21 Sep 1966.

Pileus 3–6 cm broad, convex with an incurved margin, some are obtuse and expanded to broadly umbonate with a spreading margin, glabrous, slimy viscid, dull cinnamon when fresh, drying a pale "Army Brown" overall; odor ± pungent, $FeSO_4$ no reaction on pileus (base of stipe not tested).

Lamellae moderately broad, crowded, adnate, pale cinnamon becoming "Sayal Brown" and as dried slightly redder, not beaded or spotted.

Stipe 6–9 cm long, 5–10 mm thick, equal, white and not discoloring, scurfy above, fibrillose below, pallid as dried; veil rudimentary (a thin curtain of fibrils connecting pileal margin to stipe when young).

Spores 9–12 × 5–6.5 μm, ochraceous in KOH, dextrinoid, surface minutely marbled, shape in profile view ± narrowly inequilateral, in face view narrowly to moderately boat-shaped, apex ± blunt.

Hymenium.—Basidia 4-spored, 8–10 μm broad near apex. Pleurocystidia none. Cheilocystidia filamentous-capitate, 27–57 × 8–9 × 7–11 μm and fusoid-ventricose; many 27–36 × 6×7 × 7–11 μm and not elongating prominently; subclavate to crooked or misshapen. Gill trama typical of the genus.

Lamellar and pilear tissues.—Lamellar trama typical of the genus. Cuticle of pileus a well-developed ixotrichodermium, the hyphae 1.5–2 μm diam, hyaline, smooth and clamped, sparsely branched. Hypodermium a cellular layer mixed with hyphae, ± clay color in KOH, ochraceous in Melzer's. Tramal hyphae of pileus typical of the genus.

Habit, habitat, and distribution.—Gregarious under conifers, near Kellogg, Idaho, September 21, 1966 (type, MICH).

Observations.—This species somewhat resembles *H. insigne* but does not have a scaly stipe and the spores are smaller. Also the cheilocystidia are much more versiform. It differs from *H. fastibile* in having the cuticle of the pileus an ixotrichodermium rather than an ixocutis, and in the ± cellular hypodermium.

57. Hebeloma coniferarum sp. nov.

Pileus ± 3 cm latus, plano-umbonatus, subviscidus, glaber, "Warm Sepia" vel "Verona Brown" (obscure rufobrunneus), pallescens, saepe brunneomaculatus. Contextus pallide brunneus, odor pungens, gustus mitis. Lamellae subdistantes, latae, ventricosae. Stipes ± 4 cm longus, 4 mm crassus, albofurfuraceus, deorsum sericeus et immutabiles. Velum fibrillosum, pallidum. Sporae 10–14 × 6–7 (8) μm, limoniformes, verruculosae. Cheilocystidia elongate clavata, subapicem 6–8 μm lata. Cuticula pileorum gelatinosa; pileocystidia praesentibus, cheilocydiis similibus.

Specimen typicum in Herb. Univ. Mich. conservatum est, Smith 86924; legit prope Independence Pass, Pitkin County, Colorado, 23 Jul 1976.

Pileus ± 3 cm broad, plano-umbonate, ± viscid fresh, buttons with a faint pallid fibrillose covering, soon glabrous, color at first "Warm Sepia" to "Verona Brown," paler in some at maturity, at times with watery brown spots variously disposed. Context pale watery brown fading to pinkish buff, odor slightly pungent, taste mild; $FeSO_4$ gray in the stipe; KOH no reaction on pileal cuticle.

Lamellae broad, subdistant, ventricose in age, depressed-adnate, not spotted and not beaded.

Stipe about 4 cm long and 4 mm thick, equal, tubular, watery pallid within; surface pallid to whitish, pruinose-scurfy above, silky below from pallid veil material, not discoloring in the base (as far as is known).

Spores 10–14 × 6–7 (8) μm, slightly roughened under a high-dry objective, dull cocoa colored (dull cinnamon) in KOH on fresh material; shape in profile inequilateral, in face view ovate; slowly slightly dextrinoid.

Hymenium.—Basidia 4-spored. Pleurocystidia none. Cheilocystidia elongate clavate, 6–8 μm wide at apex.

Lamellar and pilear tissues.—Lamellar trama typical for the genus. Cuticle of pileus an ixolattice and easily removed in sectioning, the hyphae about 2 μm wide and some with ends like the cheilocystidia (elongate-clavate) and 6–8 μm wide near apex; clamps present at most septa. Hypodermium cellular in tangential sections (intermixed in radial sections), ochraceous in fresh material mounted in KOH. Pileal trama typical for the genus.

Habit, habitat, and distribution.—On very rotten conifer wood, Independence Pass area, Pitkin County, Colorado, July 23, 1976, Mitchel and Smith (type, MICH).

Observations.—The failure of the base of the stipe to discolor on standing, the presence of cuticular cystidia resembling the cheilocystidia, and the watery spots on the pileus are a unique set of characters when combined with the presence of a veil. Watery spots are common among the species in subgenus *Denudata*.

58. Hebeloma kemptonae sp. nov.

Pileus 1.5–3 cm latus, convexus, demum plano-umbonatus, viscidus, glaber, ± "Pinkish Cinnamon," ad marginem pallidior; odor mitis. Lamellae luteo-pallidae demum pallide brunneae, latae, confertae, adnatae. Stipes 3–4 cm longus, 4–5 mm crassus, dissiliens, albidus, pruinose punctatus. Velum sparsum, albidum. Sporae 9–12 × 5–6 μm, punctatae leviter dextrinoideae, in "KOH" argillaceae. Basidia tetraspora. Cheilocystidia 28–57 × 4–5 × 6–12 μm, pedicellato-clavata, non agglutinata.

Specimen typicum in Herb. Univ. Mich. conservatum est, Wells and Kempton 5143; legit prope Anchorage, Alaska, 10 Aug 1971.

Pileus 1.5–3 cm broad, convex, with margin connivent, expanding to nearly flat or retaining a low umbo; color pale orange-tan (± pinkish cinnamon) on disc, paler on the margin, tacky to viscid when fresh, margin becoming slightly sulcate in some, glabrous. Context white, thin, odorless.

Lamellae moderately broad (± 5 mm), close, adnate, pale ivory becoming light brown, not beaded, edges concolorous with the faces.

Stipe 3–4 cm long, 4–5 mm broad, tending to split longitudinally, equal, whitish, white pruinose-punctate, at first thinly floccose from a rudimentary white veil, hollow.

Spore deposit about "Sayal Brown" as air-dried. Spores 9–12 ×

5–6 μm, surface minutely punctate or marbled, pale dull clay color in KOH, in Melzer's pale reddish tan (not truly dextrinoid), in face view boat-shaped to ovate, in profile inequilateral to narrowly inequilateral.

Hymenium.—Basidia 4-spored, 7–8 μm broad near apex, projecting when sporulating. Pleurocystidia none. Cheilocystidia abundant, pedicellate-clavate, the apex 6–12 μm broad (cells 28–57 × 4–5 × 6–12 μm), hyaline, thin-walled, not agglutinated.

Lamellar and pilear tissues.—Lamellar trama typical for the genus. Cuticle of pileus an ixolattice, the hyphae about 1.5 μm diam, hyaline, refractive in KOH, branched and with clamp connections. Hypodermium hyphoid, the area pale tawny in KOH, about the same in Melzer's, dextrinoid debris present in the layer. Tramal hyphae of pileus typical for the genus.

Habit, habitat, and distribution.—Along an open road, among grass and weeds bordering a mixed spruce–deciduous stand, Anchorage, Alaska, August 10, 1971, Wells and Kempton 5143 (type, MICH).

Observations.—The collectors made no mention of a veil, but the remains of a rudimentary marginal veil were ascertained from the dried specimens. This species should be compared with *H. mesophaeum* and variants which resemble it in field characters to a considerable extent. The spores and cheilocystidia, however, readily distinguish it. The cheilocystidia readily distinguish it from *H. pascuense.*

59. Hebeloma griseocanum sp. nov.

Pileus 1.5–2 cm latus, obtuse campanulatus, fibrillosus, siccus, violaceo-brunneus demum griseobrunneus. Contextus pallide brunneus, odor et gustus subraphanicus, Lamellae latae, confertae, adnatae, griseobrunneae. Stipes 2–3 cm longus, 3–3.5 mm crassus, pallidus, deorsum argillaceus. Velum copiosum, pallide griseum subannulatum. Sporae 9–12.5 × 7–9 (10) μm, dextrinoideae, in "KOH" argillaceae, subleves. Cheilocystidia 47–90 × 5–11 × 3–5 μm, ad basin ventricosa, rare cylindrica. Cuticula pileorum ixocutis est.

Specimen typicum in Herb. Univ. Mich. conservatum est, Smith 90121; legit prope Lost Man Camp, Independence Pass, Pitkin County, Colorado, 31 Aug 1979.

Pileus 1.5–2 cm broad, obtusely campanulate, dry to the touch because of a covering of veil fibrils, colors "Wood Brown" to "Benzo Brown" (dark violet-umber to dark grayish brown), margin at first fringed with remnants of the veil which is gray. Context dull brown, odor and taste pungent (raphanoid ?).

Lamellae broad, close, adnate, dull grayish brown, edges white floccose when young, neither beaded nor spotted.

Stipe 2–3 cm long, 3–3.5 mm thick, equal, pallid, pallid as dried except for the base which becomes slightly clay color, fibrillose to the fibrillose annular zone, silky above the zone. Veil copious, pale grayish and pallid as dried.

Spores 9–12.5 × 7–9 (10) μm, appearing smooth under high-dry objective, in profile view broadly inequilateral, broadly ovate in face view, dextrinoid, some with ± of an apical snout, in KOH pale clay color.

Hymenium.—Basidia 4-spored, in Melzer's when first mounted reddish but becoming orange-ochraceous on standing. Pleurocystidia none. Cheilocystidia abundant, 47–90 × 5–11 × 3–5 μm, pedicellate-ventricose and with greatly elongated neck, hyaline, thin-walled, not agglutinated; less often short and merely fusoid-ventricose, rarely cylindric.

Lamellar and pilear tissues.—Lamellar trama typical of the genus, mostly yellow revived in Melzer's. Cuticle of pileus a poorly defined ixocutis 3–6 hyphae deep and with only slightly gelatinizing walls, the hyphae 3–4 μm diam, hyaline in KOH and some with fine hyaline adhesions. Veil hyphae usually covering the cuticle. Hypodermium dark reddish brown in KOH, hyphoid, the hyphae 5–12 μm diam and with conspicuous incrustations on the walls. Clamps present.

Habit, habitat, and distribution.—Two growing together on the needle carpet under spruce, Lost Man Camp, Independence Pass, Pitkin County, Colorado, August 31, 1979 (type, MICH).

Observations.—The very broad, smooth, strongly dextrinoid spores and the grayish fibrillose pileus make a distinguishing set of characters.

60. Hebeloma pseudofastibile var. distans var. nov.

Illus. Pl. 11; figs. 7–8.

Pileus 2–4 cm latus, convexus demum late convexus, canescens, glaber, subviscidus, sordide aurantio-cinnamomeus; odor et gustus raphanicus. Lamellae subdistantes vel distantes, perlatae, ventricosae, pallide brunneolae demum subcinnamomeae. Stipes circa 4 cm longus, 8–12 mm crassus, fragilis, sursum pallidus, deorsum sordide alutaceus. Velum ± ochraceum, fibrillosum. Sporae 9–12.5 × 5.5–6.5 μm, ± leves in "KOH" ellipsoideae vel ovoideae vel phaseoliformes. Basidia tetraspora. Cheilocystidia 40–60 × 6–11 × 5–7 μm, cylindracea vel deorsum ventricosa, ad apicem subcapitata.

Specimen typicum in Herb. Univ. Mich. conservatum est, Smith 68874; legit prope McCall, Idaho, 3 Aug 1964.

Pileus 2–4 cm broad, convex with an incurved margin, expanding to broadly convex or plane; surface hoary at first from a thin veil, soon glabrous, subviscid, dingy orange-cinnamon overall and only slightly duller as dried. Context thickish, pallid brownish; odor and taste raphanoid; FeSO₄ no reaction on pileal context (not recorded for the stipe base).

Lamellae very broad and ventricose, subdistant to distant, adnate when young but soon adnexed, pallid, brownish becoming ± cinnamon, not beaded or spotted.

Stipe ± 4 cm long, 8–12 mm thick, equal, fragile, pallid above, cinnamon-buff at the base, apex silky, lower half with zones and patches from the pinkish buff veil, as dried ± cinnamon-buff or pinkish buff overall.

Spores 9–12.5 × 5.5–6.5 μm, appearing smooth in KOH, very minutely marbled in Melzer's, ellipsoid to ovoid (or some obscurely bean-shaped in profile view), nearly hyaline in KOH or in Melzer's singly, pale clay color in groups in KOH.

Hymenium.—Basidia 4-spored, 7–8 μm wide near apex. Pleurocystidia none. Cheilocystidia 40–60 × 6–11 × 5–7 μm, cylindric or ventricose near base and the neck elongated to an obtuse to subcapitate apex, hyaline, thin-walled.

Lamellar and pilear tissues.—Lamellar trama typical for the genus. Cuticle of pileus a poorly defined ixocutis, the hyphae 3–6 μm diam. Hypodermium celluar to hyphoid, The hyphal walls rusty brown in KOH and in Melzer's slightly redder, walls not conspicuously encrusted. Tramal hyphae typical for the genus. Clamps present.

Habit, habitat, and distribution.—Gregarious on sandy soil, Lake Fork Creek, McCall, Idaho, August 3, 1964 (type, MICH).

Observations.—Rarely one finds a few cheilocystidiumlike cells on the cuticle of the pileus, but it was not established whether these originated from veil hyphae or cuticular hyphae. This *Hebeloma* differs from *H. oregonense* in having smaller ellipsoid spores, in the color of the pileus, and in the apparently nondarkening stipe. It differs from *H. fastibile* in having nearly smooth spores under ordinary magnification as well as in the colored veil.

Stirps KUEHNERI

Spores dextrinoid.

KEY TO SPECIES

1. Pileus corrugated; spores ± pointed at both ends .... 61. *H. corrugatum*
1. Not as above 2
   2. Taste farinaceous; habit typically caespitose .... 62. *H. praecaespitosum*

61. Hebeloma corrugatum sp. nov.

Pileus 3–4 cm latus, late convexus demum subplanus, canescens, subviscidus, obscure luteobrunneus, ad marginem brunneogriseus, demum corrugatus. Odor et gustus mitis. Lamellae latae, ventricosae, pallidae demum brunneae. Stipes 6–8 cm longus, 2.5–4 mm crassus, deorsum demum obscure fulvus. Velum argillaceum, fibrillosum. Sporae 11–14.5 × 7–8 μm, tarde dextrinoideae, limoniformes. Cheilocystidia 43–70 × 6–9 × 4–6 μm, filiformes vel anguste fusoid-ventricosa.

Specimen typicum in Herb. Univ. Mich. conservatum est, Smith 89266; legit prope Elk Camp, Pitkin County, Colorado, 2 Sep 1978.

Pileus 3–4 cm broad, ± convex becoming plane or nearly so, surface hoary, subviscid, disc "Prout's Brown" or darker, (almost blackish brown at times), margin toned gray (± "Avellaneous"), becoming corrugated quite frequently. Context pallid, soft, odor and taste slight, FeSO$_4$ staining base of stipe dark gray.

Lamellae broad and ventricose, close to subdistant, adnate but seceding, "Verona Brown" at maturity, not beaded and not spotted, pallid when young.

Stipe 6–8 cm long, 2.5–4 mm thick at apex, equal or practically so, twisted-striate, becoming dark russet overall starting from the base and progressing upward; veil dull buff to argillaceous, thin, patches on pileus margin soon vanishing.

Spores 11–14.5 × 7–8 μm, clay color in KOH, in Melzer's slowly dextrinoid, surface marbled, shape in profile inequilateral, in face view ovate to boat-shaped (pointed at both ends).

Hymenium.—Basidia 4-spored, near apex 8–10 μm broad when sporulating. Pleurocystidia none. Cheilocystidia 43–70 × 6–9 × 4–6 μm, filiform to ventricose near the base, neck elongate and the walls often flexuous, apex obtuse and often with a refractive internal thickening in the apex, some with a secondary cross wall above the ventricose portion; many cells irregular in outline and some forked at the apex.

Lamellar and pilear tissues.—Lamellar trama typical for the genus. Cuticle of pileus a poorly defined ixocutis, the hyphae 2–3.5 μm diam, clamped, smooth to ± incrusted. Hypodermium cellular to interwoven (intermixed), with heavy pigment deposits on or in the walls, dark reddish brown in KOH. Tramal hyphae of the pileus flushed red when first mounted (revived) in Melzer's, otherwise typical for the genus.

Habit, habitat, and distribution.—Scattered under conifers, Elk Camp, Burnt Mt., Pitkin County, Colorado, September 2, 1978 (type, MICH). See also Smith 87089 and 89078.

Observations.—Collections of this species in which the pileus is not, or only slightly, wrinkled can be identified by the spores many of which are ± pointed at both ends. This along with the dingy tan veil, stipe becoming dark russet to date brown, lack of a distinctive odor and taste, and variability of the cheilocystidia as described characterize the species.

62. Hebeloma praecaespitosum sp. nov.

Illus. Pl. 5*A;* figs. 32–33, 51, and 53–54.

Pileus 1.5–4 cm latus, obtusus vel convexus, demum subplanus, udus, hygrophanus, sparse fibrillosus demum glaber, "Cinnamon-Brown" demum fulvus; odor acidulus, gustus farinaceus. Lamellae pallide cinnamomeae demum obscure "Sayal Brown," latae, confertae. Stipes 4–9 cm longus, 4–7 mm crassus, pallidus, tarde brunnescens, dissilliens. Velum fibrillosum pallide ochraceum. Sporae 10–13 × 6.5–7.5 μm, in "KOH" argillaceae, subleves, ± dextrinoideae inequilaterales. Basidia tetraspora. Pleurocystidia et cheilocystidia similis (36) 48–75 × (6) 8–12 × 3.5–5 μm, fusoide ventricosa vel subfilamentosa.

Specimen typicum in Herb. Univ. Mich. conservatum est, Smith 89627; legit prope Snowmass Village, Pitkin County, Colorado, 3 Aug 1979, Ammirati and Smith.

Pileus 1.5–4 cm broad, obtuse to convex, expanding to nearly plane, margin incurved at very first, surface moist and hygrophanous at times with a few veil flecks on the disc but mostly the veil remnants distributed along the margin evenly or in patches; color when moist "Cinnamon-Brown," becoming fulvous and finally dull cinnamon ("Sayal Brown"), when dried and in the herbarium near "Warm Sepia"—a dark reddish brown. Context watery brown, pallid when faded, odor sharp (acidulous), taste farinaceous; $FeSO_4$ quickly staining base of stipe olive-fuscous.

Lamellae broad, close, adnate-seceding, pale "Cinnamon" when young, near "Sayal Brown" when mature (a dull cinnamon color), not beaded and not spotted.

Stipe 4–9 cm long, 4–7 mm thick, equal, pallid at first, soon flushed brown starting at base and spreading upward to the apex, pallid again as dried, readily splitting lengthwise, surface ragged from remnants of the veil; veil fibrillose, with a thin buff-colored outer layer and a thicker inner pallid layer.

Spores 10–13 × 6.5–7.5 μm, appearing smooth and pale clay color in KOH, in Melzer's weakly ornamented and moderately dextrinoid; shape in face view elliptic to ovate to broadly subfusoid, in profile mostly somewhat inequilateral.

Hymenium.—Basidia 4-spored, 8–12 μm broad near apex, hyaline globules present in some basidia. Pleurocystidia similar to cheilocystidia and present mostly near the gill edge. Cheilocystidia (36) 48–75 × (6) 8–12 × 3.5–5 μm, narrowly fusoid-ventricose to subcylindric, neck often flexuous, rarely branched, not agglutinated, some remaining basidiolelike.

Lamellar and pilear tissues.—Lamellar trama typical of the genus except for the red flush which develops in Melzer's especially near the base of the hymenium. Cuticle of pileus a well-defined ixocutis of hyaline hyphae 2–3.5 μm diam with gelatinizing walls, clamps present at the septa (note: brown narrow tubular hyphae often in a layer above the ixocutis are here interpreted as veil remnants). Hypodermium appearing cellular in tangential sections, intermixed in radial sections, dull tawny at first in KOH but fading to ochraceous, walls not significantly encrusted as seen in KOH, in Melzer's some dextrinoid debris present and the incrustations more apparent. Pileus trama of radial interwoven hyphae, the cells inflated and the walls mostly smooth. Clamps present on all tissues of the fruiting body.

Habit, habitat, and distribution.—Densely caespitose under Englemann spruce above Snowmass Village, Pitkin County, Colorado, alt. 9,200 ft, August 3, 1979 (type, MICH.).

Observations.—Sections of fresh material showed a variation in spore size of 9–11 × 5–6 μm to 10–13 × 6.5–7.5 μm between fruiting bodies in a single cluster. The diagnostic characters for the species are: (1) the two-toned fibrillose veil, (2) stipe darkening overall, (3) dark brown fresh pileus, (4) the sharp odor and ± farinaceous taste, (5) dextrinoid spores, (6) reddish flush of the lamellar trama in Melzer's (on dried material), (7) the nonagglutinating cheilocystidia and (8) the ± inequilateral shape of the spores in profile view.

63. Hebeloma fuscostipes sp. nov.

Pileus 1.5–2 cm latus, convexus demum expansus, ad marginem griseofibrillosus, glabrescens, subviscidus, triste rufobrunneus vel ad marginem pallidior; odor et gustus mitis. Lamellae latae (ventricosae), confertae, adnatae, sordide cinnamomeae. Stipes 2–3 cm longus, 1.5–2.5 mm crassus, aequalis, fragilis, deorsum tactu triste brunneus. Velum griseum, fibrillosum. Sporae 11–14 × 6.5–7.5 μm, dextrinoideae. Cheilocystidia 38–65 × 5–11 × 4–6 μm, elongate fusoid-ventricosa, rare filamentosa, 36–70 × 4–6 μm.

Specimen typicum in Herb. Univ. Mich. conservatum est; Smith 90501; legit prope Elk Camp, Burnt Mt., Pitkin County, Colorado, 2 Aug 1980.

Pileus 1.5–2 cm broad, convex becoming plane, margin at first decorated with bits of the fibrillose grayish veil, glabrous finally, subviscid to viscid when fresh, color "Warm Sepia" over disc (dark reddish brown), margin paler (± "Verona Brown"), opaque at all times. Context watery brownish fading to pallid, odor and taste mild, $FeSO_4$ staining base of stipe black.

Lamellae broad (ventricose), close, adnate, ± "Verona Brown," to dull cinnamon at maturity (± "Sayal Brown"), neither spotted nor beaded.

Stipe 2–3 cm long, 1.5–2.5 mm thick, equal, fragile, becoming dark brown at base and then upward; surface silky-fibrillose and grayish, ± scurfy from pruina and veil remnants combined; staining brown when handled.

Spores 11–14 × 6.5–7.5 μm, ± inequilateral in profile, ovate in face view, clay color revived in KOH, dextrinoid, punctate as viewed in Melzer's reagent.

Hymenium.—Basidia 4-spored, 7–10 μm broad near apex; hymenium red in Melzer's. Pleurocystidia none. Cheilocystidia mostly 38–65 × 5–11 × 4–6 μm, narrowly fusoid-ventricose, apex obtuse to slightly enlarged; some filamentous and 36–70 × 4–6 μm, as revived in KOH ochraceous to ± cinnamon-buff in KOH but hyaline after standing for about 20 minutes.

Lamellar and pilear tissues.—Lamellar trama typical for the genus except for the red color in mounts in Melzer's reagent—but this soon fading to orange. Cuticle of pileus an ixocutis, the hyphae 2–4 mm diam, clamped, sparingly branched. Hypodermium distinctly hyphoid and in KOH the hyphal walls heavily encrusted with dark brown material, in Melzer's the incrustations not conspicuous. Tramal hyphae ± typical of the genus but red in Melzer's at first but quickly fading.

Habit, habitat, and distribution.—Scattered under spruce, near Burnt Mt., Pitkin County, Colorado, August 2, 1980 (type, MICH).

Observations.—The very slender fragile stipe readily staining brown when handled, lack of both an odor and distinctive taste, and the conspicuous incrustations on the hypodermal hyphae, are distinctive as a combination. The species appears similar to *H. corrugatum*, but in face view the spores are hardly boat-shaped, the veil is grayish (not clay color), the hypodermium is clearly hyphoid (not at all "cellular"), and the stipe of *H. corrugatum* is not known to stain brown from handling.

64. Hebeloma subumbrinum sp. nov.

Pileus 1–2.5 cm latus, convexus vel late expansus, viscidus, canescens "Wood Brown" vel "Avellaneous," ad marginem fibrillosus; odor raphanicus. Lamellae "Avellaneous" demum incarnato-cinnamomeae, latae, confertae, adnatae. Stipes 3–5 cm longus, 3–4 mm crassus, obscure brunneus, sursum pallidus; mycelium subroseum. Sporae 10–12.5 × 6–7.5 μm, dextrinoideae. Cheilocystidia 38–76 × 9–12 (21) × 6–7 μm, fusoid-ventricosa.

Specimen typicum in Herb. Univ. Mich. conservatum est, Smith 88979; legit prope Independence Pass, Colorado, 14 Aug 1978.

Pileus 1–2.5 cm broad, obtuse to convex becoming broadly convex, surface thinly viscid, at first with a hoary coating of veil fibrils at least near the margin, color beneath the veil remnants "Wood Brown" (brownish gray). Context watery avellaneous fading to pallid, odor raphanoid-pungent; $FeSO_4$ staining base of stipe quickly to olive-black.

Lamellae broad, close, adnate, vinaceous-buff to avellaneous becoming dingy pinkish cinnamon, edges not beaded.

Stipe 3–5 cm long, 3–4 mm thick, dark brown from base upward, pallid overall at first, with a pinkish buff pad of mycelium around basal area; veil pallid and fibrillose, leaving thin patches variously disposed over the surface.

Spores 10–12.5 × 6–7.5 μm, clay color to dull ochraceous tawny in KOH, slowly bay-red in Melzer's (dextrinoid), surface minutely marbled; shape in profile ± inequilateral, in face view ovate to boat-shaped.

Hymenium.—Basidia 4-spored, 8–10 μm broad near apex when sporulating. Pleurocystidia none. Cheilocystidia fusoid-ventricose to filamentous, 38–76 × 9–12 (21) × 6–7 μm, ventricose near base and with an elongating often flexuous neck, apex merely obtuse. Secondary septa not observed in the neck, no forking observed, the walls of the neck slightly refractive in KOH but the cystidia were not observed to agglutinate.

Lamellar and pilear tissues.—Lamellar trama typical of the genus; hymenium orange-ochraceous in Melzer's. Cuticle of pileus a poorly defined ixocutis of hyphae 3–4 μm diam and with ± refractive walls (hyphae of the nongelatinous veil often seen on the surface). Hypodermium hyphoid, rusty brown in KOH and reddish brown in Melzer's, hyphae encrusted, dextrinoid debris ± copious. Tramal hyphae typical of the genus (no red content in Melzer's). Clamps present.

Habit, habitat, and distribution.—Solitary to scattered on conifer debris, Lincoln Creek, Pitkin County, Colorado, August 14, 1978 (type, MICH).

Observations.—Note the scattering of greatly inflated cheilocystidia. The species differs from *H. obscurum* in having pinkish buff basal mycelium and broader cheilocystidia.

65. Hebeloma occidentale sp. nov.

Illus. Fig. 52.

Pileus 2–5 cm latus, "Warm Sepia" deinde "Ochraceous-Tawny," ad marginem cano-cineraceus, viscidus. Contextus pallide griseobrun-

neus; odor et gustus mitis. Lamellae pallide brunneae demum argillaceae, latae, confertae. Stipes 4–7 cm longus, 4–7 mm crassus, pallidus,
cavus, deorsum brunnescens, dense albofibrillosus. Velum pallidum,
fibrillosum. Sporae 9–12 × 5.5–7.5 (9) μm, inequilaterales, dextrinoideae, subleves, pallide argillaceae, in "KOH." Basidia 28–32 × 6–8
μm, tetraspora. Pleurocystidia nulla. Cheilocystidia 30–62 × 6–12 × 4–
9 μm, deorsum ventricosa, ad apicem ± capitata.

Specimen typicum in Herb. Univ. Mich. conservatum est, Smith
19106; legit Mt. Hood National Forest, Oregon, 25 Sep 1944.

Pileus 2–5 cm broad, becoming plane or the margin slightly wavy
and incurved, not umbonate, viscid, at first "Warm Sepia" on the disc,
hygrophanous, fading to "Ochraceous-Tawny," the marginal area
grayish fibrillose or hoary; context watery grayish brown, pallid when
faded; odor and taste mild or none.

Lamellae broad, close, adnate, pallid brown becoming nearly
clay color, with two tiers of lamellulae, edges white-fimbriate.

Stipe 4–7 cm long, 4–7 mm thick, pallid, base becoming bister in
age or where bruised, equal, hollow, fragile, densely white-fibrillose, in
age silky, glabrescent; veil of pallid fibrils forming an evanescent zone.

Spores 9–12 × 5.5–7.5 (9) μm, inequilateral in profile, subovate
to subelliptic in face view, wall 0.4–0.5 μm thick, minutely rugulose,
pale yellowish brown in KOH, dextrinoid.

Hymenium.—Basidia 28–32 × 6–8 μm, 4-spored. Pleurocystidia
none. Cheilocystidia 30–62 × 6–12 × 4–9 μm, ventricose near base,
apex ± capitate.

Lamellar and pilear tissues.—Lamellar trama typical of the genus.
Cuticle of pileus an ixocutis, the layer ± 60 μm thick. Hypodermium a
zone of brown cells (in KOH). Stipe cuticle of dry brown repent hyphae, caulocystidia scattered, similar to cheilocystidia, occasionally
one seen with a clamp.

Habit, habitat, and distribution.—In muck soil under *Alnus*, Mt.
Hood National Forest, Cascade Mts., Oregon, September 25, 1944
(type, MICH).

Observations.—The pallid veil, dark red-brown pileus fading
markedly to ochraceous-tawny, mild odor and taste, brunnescent stipe,
wide spores, narrow basidia and habitat under alder are significant.
The spores readily distinguish it from variants of *H. mesophaeum.*

66. Hebeloma hesleri sp. nov.

Pileus 3–5 cm latus, convexus demum late convexus vel subplanus, subviscidus, sordide cinnamomeus vel dilute rufobrunneus, ad
marginem griseobrunneus et fibrillosus; velum griseum. Contextus di-

lute brunneus vel pallidus, odor pungens (subraphaninus). Lamellae adnatae, pallidae demum subalutaceae vel subfulvae, latae, confertae. Stipes 4–6 cm longus, 5–9 mm crassus, pallidus, deorsum triste fulvus, sursum pruinosus. Sporae 9–13 × 5–6 μm, inequilaterales, tarde dextrinoideae, minute rugulosae vel subleves. Basidia tetraspora. Cheilocystidia versiforma: (1) (20) 34–85 × 7–10 μm, cylindrica vel elongate clavata; (2) 34–85 × 7–10 μm, fusoide ventricosa.

Specimen typicum in Herb. Denver Bot. Garden conservatum est, Smith 86922; legit Independence Pass, Colorado, 27 Jul 1976, D. H. Mitchel.

Pileus 3–5 cm broad, obtuse to convex, expanding to nearly plane, disc ± dull cinnamon ("Sayal Brown"), becoming "Verona Brown" (dull reddish brown), margin paler and grayish from a thin coating of veil fibrils, in age glabrescent and colors paler, subviscid when moist. Context pallid to watery brown, odor pungent to raphanoid, taste mild; FeSO₄ staining base of stipe dull olive; KOH not staining the tissues.

Lamellae broad, close, adnate, pallid to pinkish buff, becoming dull cinnamon, not spotting, thin, edges even.

Stipe 4–6 cm long, 5–9 mm thick, pallid within, becoming dull dark rusty brown from the base up, surface thinly fibrillose, white-pruinose to scurfy above, silky over the remainder at maturity, equal, soon hollow. Veil remnants as evanescent pallid fibrils on the stipe.

Spores 9–13 × 5–6 μm, inequilateral in profile, subelliptic in face view, varying to ovate, minutely marbled (or appearing ± smooth), yellowish brown in KOH; slowly dextrinoid.

Hymenium.—Basidia 4-spored. Cheilocystidia of two types: (1) (20) 34–85 × 7–10 μm, cylindric to elongate-clavate, and (2), 34–85 × 7–10 × 4–6 μm, fusoid-ventricose with a long slender neck and often a long pedicel.

Lamellar and pilear tissues.—Lamellar trama typical for the genus. Cuticle of pileus an ixocutis. Hypodermium cellular. Stipe cuticle of appressed dry hyphae. Caulocystidia in tufts, similar to the cheilocystidia. Clamp connections present.

Habit, habitat, and distribution.—On soil, gregarious under conifers, Independence Pass, Pitkin County, Colorado, D. H. Mitchel (Smith 86922), July 27, 1976 (type, DBG).

Observations.—We do not place taxonomic emphasis on the odor being raphanoid and the taste mild, though admittedly the situation is peculiar. The diagnostic features are the gray veil, two types of cheilocystidia (and the fact that they become very elongated on mature specimens), the distinctly dark rusty brown color change of the stipe, and the cellular hypodermium.

67. Hebeloma obscurum sp. nov.

Pileus (1.5) 2.5–4.5 cm latus, late convexus vel planus, canescens, ad marginem fibrillose squamulosus, leviter viscidus. Contextus mollis, odor pungens; gustus mitis. Lamellae latae, subdistantes, pallidae demum argillaceo-cinnamomeae. Stipes 3–6 cm longus, 3–6 mm crassus, deorsum sordide brunneus (spadiceus), sursum fibrillosus, subannulatus. Sporae 10–13 × 5.5–7 μm, dextrinoideae, inequilaterales. Cheilocystidia 37–64 × 7–10 × 3–5 × 4–6 μm, fusoid-ventricosa.

Specimen typicum in Herb. Univ. Mich. conservatum est, H. D. Thiers and A. H. Smith 88923; legit prope Mormon Creek, Pitkin County, Colorado, 9 Aug 1978.

Pileus (1.5) 2.5–4.5 cm broad, convex becoming plane, surface hoary from a thin coating of veil fibrils giving it a hoary overtone, margin decorated with thin patches of grayish veil fibrils, ground color about "Benzo Brown" beneath the veil, subviscid but soon dry, opaque. Context dingy watery brown, soft, dingy buff when faded; odor pungent; taste mild; $FeSO_4$ staining base of stipe blackish.

Lamellae broad, subdistant to close, adnate, pallid when young, becoming avellaneous and finally "Sayal Brown" from the spores, edges not beaded.

Stipe 3–6 cm long, 3–6 mm thick, equal, soon bister below, grayish silky above, veil-line ± persistent, surface often twisted-striate.

Spores 10–13 × 5.5–7 μm, clay color in KOH, slowly dextrinoid, marbled to appearing ± smooth, in profile view mostly inequilateral, ovate in face view, apex blunt and not projecting as a snout.

Hymenium.—Basidia 4-spored, 26–34 × 7–10 μm. Pleurocystidia none. Cheilocystidia 37–64 × 7–10 × 3–5 × 4–6 μm, ventricose near the base and the neck greatly elongated but lacking secondary septa and no forking observed, some cheilocystidia filamentous and flexuous, none distinctly enlarged at the apex.

Lamellar and pilear tissues.—Lamellar trama typical for the genus. Cuticle of pileus a distinct ixocutis of hyphae 2.5–4 μm diam, mostly hyaline, many with fine incrustations or roughenings on the wall. Hypodermium cellular, ochraceous in KOH on fresh material, dark rusty brown as revived, pigment deposits and incrustations numerous, dextrinoid debris present in the layer and in the ixocutis. Tramal hyphae typical for the genus but cells more inflated than usual in old specimens. Clamp connections present.

Habit, habitat, and distribution.—Gregarious under lodgepole pine and spruce, Fryingpan River near Mormon Creek, Pitkin County, Colorado, August 9, 1979 (type, MICH).

Observations.—The dark violaceous brown pileus, intense $FeSO_4$

reaction, rather heavy grayish veil, and dextrinoid debris in the cuticle of the pileus make a distinctive combination of characters. *H. nigellum* resembles it in some respects, but has a raphanoid then bitter taste, and the pileus lacks a violaceous tone. *H. kuehneri* is also close but in *H. obscurum* the gills are pallid then avellaneous, and the veil is gray.

68. Hebeloma griseovelatum sp. nov.

Pileus 2–2.5 cm latus, demum planus vel obscure umbonatus, umbrinorufus deinde sordide fulvus, subviscidus, leviter fibrillosus, canescens; contextus tenuis, pallide argillaceus, odor et gustus mitis. Lamellae adnatae, confertae, angustae demum sublatae. Stipes 3–4 cm longus, 3–5 mm crassus, deorsum fulvus, griseofibrillosus. Velum griseum, fibrillosum. Sporae 9–12 × 4.5–6 μm, inequilaterales, minute rugulosae. Basidia tetraspora. Pleurocystidia nulla. Cheilocystidia fusoide ventricosa vel subclavata, 42–70 × 4–10 μm, et 27 × 9 μm.
Specimen typicum in Herb. Univ. Mich. conservatum est, Smith 86929; legit prope Independence Pass, Pitkin County, Colorado, 27 Jul 1976.

Pileus 2–2.5 cm broad, obtuse, the margin inrolled, expanding to plane or with an obscure umbo, color evenly pale "Warm Sepia," fading to dull tawny, margin inrolled and long remaining so; slightly viscid, fibrillose and appearing hoary. Context thin, more or less pliant-firm, pale buff; odor and taste not distinctive. FeSO$_4$ olive-fuscous in cap and stipe; with KOH no reaction.

Lamellae narrow becoming only moderately broad, moderately close, broadly adnate or with a decurrent tooth, grayish (near avellaneous) then more or less "Verona Brown," not beaded and not staining.

Stipe 3–4 cm long, 3–5 mm thick, dark rusty brown in the base and changing to brown upward gradually, firm, soon hollow, surface grayish fibrillose, apex ± naked, some long-striate in age. Veil fibrils grayish.

Spores 9–12 × 4.5–6 μm, inequilateral in profile, elliptic in face view, minutely rugulose, very pale yellowish in KOH.

Hymenium.—Basidia 4-spored. Pleurocystidia none. Cheilocystidia of two types: *a*) 42–70 × 4–10 μm and fusoid-ventricose, and *b*) 27 × 9 μm and ± clavate.

Lamellar and pilear tissues.—Lamellar trama typical of the genus. Cuticle of pileus a thin layer of gelatinous hyphae (a rudimentary ixocutis). Hypodermium cellular. Stipe cuticle dry, hyphae repent. Caulocystidia as terminal elements on epicuticular hyphae of stipe and similar to the cheilocystidia.

Habit, habitat, and distribution.—On soil, gregarious, under conifers, Independence Pass, Pitkin County, Colorado, July 27, 1976 (type, MICH).

Observations.—The two types of cheilocystidia are not as valuable a feature here as one might expect because some clavate cells appear on or near the gill edge in many of the small veiled species. *Hebeloma sterlingii* (Pk.) Murrill is close to *H. griseovelatum* but distinguished by a farinaceous taste and the interior of the stipe is said to be bay-red. *H. griseovelatum* is also close to *H. testaceum* sensu Bruchet but differs in not having a distinctive odor and taste. Also, its pileus is scarcely viscid and the veil is distinctly grayish.

69. Hebeloma piceicola sp. nov.

Illus. Pl. 6*A;* figs. 41–42..

Pileus 1–2 cm latus, demum plano-umbonatus, canescens, ad marginem fibrillosus, glabrescens, griseobrunneus. Odor et gustus raphanicus. Lamellae latae, confertae, adnatae, vinaceopallidae demum subcinnamomeae. Stipes 2.5–3.5 cm longus, 3–4.5 mm crassus, "Vinaceous-Buff," deorsum sordide brunneus, fibrillosus. Sporae 10–12.5 × 6–7 μm, ± dextrinoideae. Cheilocystidia 43–58 × 4–6 μm, subcapitata.

Specimen typicum in Herb. Univ. Mich. conservatum est, Smith 89135; legit prope Elk Camp, Burnt Mt., Pitkin County, Colorado, 25 Aug 1978.

Pileus 1–2 cm broad, obtuse then expanding to plano-umbonate, hoary at first from fibrils of an outer veil, patches of veil fibrils along the margin persisting for some time, ground color gray-brown ("Wood Brown"), opaque at all stages. Context dark watery brown, odor and taste ± raphanoid; $FeSO_4$ staining the base of the stipe olive-green to blackish.

Lamellae vinaceous-buff, becoming a pale "Verona Brown," broad, adnate to adnexed, close but not crowded, edges neither beaded nor spotted.

Stipe 2.5–3.5 cm long, 3–4.5 mm thick, equal, ± vinaceous-buff above, becoming bister in the base and the change progressing upward, surface pallid from the fibrillose veil, the fibrils finally evanescent.

Spores 10–12.5 × 6–7 μm, surface minutely asperulate in Melzer's, ± dextrinoid, ± clay color and smooth in KOH, fading to ± hyaline, shape in profile inequilateral, in face view ovate.

Hymenium.—Basidia 4-spored, 8–10 μm broad near apex. Pleurocystidia none. Cheilocystidia ± cylindric and 43–58 × 4–6 μm, only slightly enlarged at apex and some with an annular zone a short distance back from the apex (as revived), ± refractive in KOH and hyaline.

Lamellar and pilear tissues.—Lamellar trama typical for the genus. Cuticle of pileus a thin ixocutis, hyphae 2–3.5 μm diam, the layer often obscured by overlying veil hyphae. Hypodermium hyphoid, in KOH with heavy reddish brown pigmented deposits on the wall; color in Melzer's a brighter reddish brown. Tramal hyphae typical for the genus. Veil hyphae tubular, 3–6 μm diam, clamp connections present, hyaline in KOH.

Habit, habitat, and distribution.—Scattered to gregarious on swampy ground (muck) under spruce, Elk Camp, Pitkin County, Colorado, August 25, 1978 (type, MICH).

Observations.—The pallid veil, strongly inequilateral medium-sized spores, canescent pileus, and raphanoid odor and taste indicate a relationship to *H. idahoense* from which it differs mainly in the characters of the cheilocystidia.

70. Hebeloma pseudofastibile sp. nov.
var. pseudofastibile

Pileus 3.5–4 cm latus, convexus demum late expansus, glaber vel ad marginem albofibrillosus, argillaceus, subviscidus, demum subrimosus. Contextus brunneolus, odor et gustus raphanicus. Lamellae latae, confertae, sordide cinnamomeae. Stipes 6–9 cm longus, 8–9 mm crassus, fragilis, dissiliens, ad basin fulvescens. Vellum pallidum demum subochraceum. Sporae 9–12 (13) × 6–7 μm, inequilaterales, leviter dextrinoideae, subleves. Basidia tetraspora. Cheilocystidia 40–80 × 4–9 × 3–5 μm, hyalina; vel filamentosa, 3.5–5 μm diam ad apicem obtusa.

Specimen typicum in Herb. Univ. Mich. conservatum est, Smith 90425; legit prope Snowmass Village, Pitkin County, Colorado, 30 Jul 1980.

Pileus 3.5–4 cm broad, broadly convex, glabrous but at first with white patches of fibrils along the margin, pinkish tan to clay color over the disc, margin paler, scarcely viscid, cuticle tending to become rimose. Context brownish, thin but firm, odor and taste raphanoid, $FeSO_4$ staining base of stipe blackish.

Lamellae broad, close, adnate, thin, more or less "Sayal Brown" at maturity (dull cocoa-brown), beaded with hyaline drops at first, edges serrulate.

Stipe 6–9 cm long, 8–9 mm thick at apex, equal, fragile, splitting lengthwise fairly readily, becoming dull rusty brown in the base and then upward; surface pallid-fibrillose and over this layer occurs remnants of the veil as pallid patches which gradually stain (become) pinkish buff in about 2 hours.

Spores 9–12 (13) × 6–7 μm, inequilateral in profile view, in face view ovate to subfusiform, weakly dextrinoid, obscurely ornamented (punctate in Melzer's reagent).

Hymenium.—Basidia 7–9 μm broad, clavate, reddish in Melzer's but soon orange-ochraceous, content (as seen in KOH) including hyaline globules in the apex. Pleurocystidia present only near the gill edge, similar to cheilocystidia. Cheilocystidia fusoid ventricose with greatly elongated often wavy necks and apex obtuse to weakly enlarged, 40–80 × 4–9 × 3–5 μm, content hyaline in KOH, many shaped like a hockey stick; also a fair number filamentous to cylindric and 3.5–5 μm diam, apex not significantly enlarged.

Lamellar and pilear tissues.—Lamellar trama typical of the genus but at first dull red in Melzer's, soon fading out to ± ochraceous-orange; hyphal cells elongate and considerably enlarged (8–18 μm ±). Cuticle of pileus an ixocutis, hyphae tubular and 2–5 μm broad, clamped. Hypoderm ± hyphoid to cellular, some incrustations present on the walls, reddish brown in KOH. Tramal hyphae typical of the genus, in Melzer's merely yellowish.

Habit, habitat, and distribution.—On a stream bank under conifers and willows, Snowmass Village, Pitkin County, Colorado, July 30, 1980 (type, MICH).

Observations.—This species is closely related to *H. fastibile* but differs in the Melzer's reaction of the hymenium and spores, in the brunnescent stipe and, apparently, in the manner in which the veil changes color over the darkening portion of the stipe.

71. **Hebeloma remyi** Bruchet

Bull. Soc. Linn. de Lyon, 39 année: 20. Jun 1970

Illus. Bruchet, l.c., fig. 14. Pl. 6*B*.

Pileus 2–4.5 cm broad, broadly convex to plane in age, disc at times shallowly depressed; margin at first decurved and inrolled, glabrous except for brownish patches of the veil, canescent at first when moist; color cinnamon-brown to watery tan on disc, margin pinkish buff to grayer, opaque at all times (when young some pilei ± violaceous-umber to drab). Context watery brown, odor and taste distinctly of radish; FeSO$_4$ staining base of stipe olive-fuscous.

Lamellae broad, subdistant, ventricose, adnate, ± avellaneous becoming dull cinnamon, edges eroded but not beaded or spotted.

Stipe 3–7 cm long, 3–5 mm thick, equal, ± bister to snuff brown below, whitish above, fibrillose from veil remnants; veil white at first on button stages, pale brownish in age.

Spore deposit clay color. Spores 11–14 × 6.5–8 μm, ovate to subfusoid in face view, ± inequilateral in profile, ± smooth in KOH, faintly ornamented (punctate to rugulose) in Melzer's, slowly dextrinoid, apex ± snoutlike in some.

Hymenium.—Basidia 4-spored, 9–12 μm broad, clavate, content conspicuous because of droplets (in KOH mounts). Pleurocystidia none. Cheilocystidia basically fusoid-ventricose, the apex obtuse, 30–68 × 8–10 × 5–7 μm (for the ventricose cystidia), some ± filamentous and 3–6 μm diam, thin-walled and not agglutinated.

Lamellar and pilear tissues.—Lamellar trama typical for the genus (no rose tones in mounts in Melzer's) and no dextrinoid debris noted. Cuticle of pileus a poorly defined ixocutis of hyphae 2–3.5 μm diam, tubular and walls ± gelatinized, the layer 3–5 hyphae deep. Hypodermium cellular, the cells greatly inflated, walls ochraceous-brown in KOH and with heavy incrustations dark rusty brown, no dextrinoid debris noted. Hyphae of pilear trama ochraceous-hyaline in KOH, smooth, cells of various diameters. Clamp connections present.

Habit, habitat, and distribution.—Gregarious under spruce, West Branch of Brush Creek, Burnt Mt., Pitkin County, Colorado, August 22, 1979, Smith 89846 (MICH); also 89888.

Observations.—The diagnostic features of the American collections are: (1) the white veil on button stages and the brownish patches on the mature or old pilei, (2) the large dextrinoid spores inequilateral in profile view, (3) the weak ornamentation of the spores considering their size, (4) the essentially fusoid-ventricose cheilocystidia, (5) the broadly ventricose subdistant gills, (6) the raphanoid odor and taste, and (7) the lack of red tones in the pileus as revived in Melzer's. Although spruce trees were present in the habitat there were also many shrubs including alder and willow. As to habitat, all that can be said to date of *H. remyi* here in North America is that it is a montane species.

72. **Hebeloma marginatulum** (Favre) Bruchet

Bull. Soc. Linn. de Lyon, 39 année: 43. Jun 1970

var. **marginatulum**

Pileus 2–3.5 cm broad, broadly convex to nearly plane, edge usually remaining decurved, at first hoary from remains of a thin fibrillose veil, ground color dark vinaceous-brown ("Natal Brown") overall, margin remaining coated with veil remnants for some time, as dried about "Army Brown" overall; odor and taste distinctly raphanoid.

Lamellae broad, close, adnate, pallid when young, rusty cinnamon at maturity, not beaded or spotted.

Stipe short (2–3 cm long), 3–4 mm thick, silky, pallid and with only a faint veil-line and this soon evanescent, brunnescent and darkening from below upward, as dried brownish pallid overall.

Spores 10–13 (14.5) × 6.5–7.5 μm, apparently smooth under high-dry objective but in Melzer's under an oil-immersion lens faintly mottled; clay color in KOH, dextrinoid, in profile ± inequilateral, in face view ovate.

Hymenium.—Basidia 4-spored, 9–12 μm wide. Pleurocystidia none. Cheilocystidia (37) 48–70 × 5–10 × 4–6 (16) μm; some elongated and cylindric, or walls flexuous, some narrowly fusoid-ventricose, some ± elongate-clavate, all types hyaline in KOH.

Lamellar and pilear tissues.—Lamellar trama typical for the genus. Cuticle of pileus an ixolattice, the hyphae 2–4 (5) μm diam, hyaline to yellowish in KOH. Hypodermium the intermediate type, reddish rusty in KOH and with conspicuous colored incrustations, the layer bay-red in Melzer's. Tramal hyphae of pileus flushed red in Melzer's but soon fading. Clamps present.

Habit, habitat, and distribution.—Under alder, South Fork, Lake Fork Creek, McCall, Idaho, September 10, 1956; Smith 53299; Colorado, Smith 87552, 88911.

Observations.—On the basis of the top-heavy appearance of the basidiocarps, the dark-colored pilei, the large and ± inequilateral spores, the long cheilocystidia with flexuous walls and the poorly developed ixocutis of the pileus, we assign our collections to *H. marginatulum.* Bruchet's figures 1–2 express the aspect of the North American fruiting bodies very well. Apparently this is a very rare species in North America.

72a. Hebeloma marginatulum var. fallax var. nov.

Illus. Figs. 45–46.

Pileus 2.5–4 cm latus, plano-convexus, ad marginem griseobrunneus, sparsim fibrillosus, ad centrum obscure vinaceobrunneus, glabrescens. Contextus brunneolus demum pallidus, odor et sapor raphaninus. Lamellae adnatae, latae, demum subdistantes, brunneolae demum obscure cinnamomeae. Stipes 3–7 cm longus, (2) 3–5 mm crassus, dissiliens, deorsum brunneolus. Velum sparsim, pallidum vel cinereum. Sporae 10–15 × 6.5–8 μm, inequilaterales, pallide cinnamomeae (cum "KOH"), dextrinoideae, subleves. Basidia tetraspora. Pleurocystidia nulla. Cheilocystidia conspicua, cylindrica, vel ventricosa ad basin, elongata, 40–90 × 5–8 × 3–5 μm.

Specimen typicum in Herb. Univ. Mich. conservatum est, Smith 89724; legit prope Snowmass Village, Pitkin County, Colorado, 12 Aug 1979.

Pileus 2.5–4 cm broad, plano-convex, the margin dull grayish brown and thinly coated with grayish buff fibrils from the veil, disc dull reddish brown and dried evenly ± "Army Brown" (vinaceous-brown) overall, glabrescent. Context with a raphanoid odor and taste, brownish fading to pallid, $FeSO_4$ staining stipe base olive.

Lamellae broad, close to subdistant, adnate, pallid dull brown becoming wood brown and finally dull cinnamon ("Sayal Brown"), not beaded.

Stipe 3–7 cm long, (2) 3–5 mm thick, equal to slightly enlarged below, pallid and thinly fibrillose, as dried evenly brownish overall, changing to brown from the base upward when fresh, very fragile and readily splitting lengthwise. Veil cortinate and thin, remains ± evenly dispersed, buff colored where in distinct patches.

Spore deposit (on stipes) dull cinnamon. Spores 10–15 × 6.5–8 μm, ovate in face view, inequilateral in profile, pale cinnamon as revived in KOH, dextrinoid, appearing smooth in outline, with a large central globule as seen mounted in KOH.

Hymenium.—Basidia 4-spored. Pleurocystidia none. Cheilocystidia conspicuous, cylindric to ventricose-elongate, 40–90 × 5–8 × 3–5 μm, basal area in some with a pale brown homogeneous content (or color located in the wall ?) not becoming agglutinated as far as observed.

Lamellar and pilear tissues.—Lamellar trama with a slight rose-red reaction as mounted in Melzer's but soon fading and the hymenium ± ochraceous or orange-ochraceous. Cuticle of pileus a poorly defined ixocutis. Hypodermium of loosely interwoven hyphae 12–20 μm diam, cells short to long, in Melzer's the heavy incrustations highly pigmented (bay-red), in KOH duller but nevertheless colored. Clamps present. Tramal hyphae flushed red in Melzer's but quickly fading.

Habit, habitat, and distribution.—Gregarious on humus under brush and aspen with conifers nearby; near Snowmass Village, Pitkin County, Colorado, August 12, 1979 (type, MICH).

Observations.—The pale cinnamon color of the spores in KOH, the loosely organized hypodermium, the buff cortina and in this variety in particular, the raphanoid odor and taste, are critical characters. The cheilocystidia in some KOH mounts showed interior apical thickenings which dissolved in KOH in approximately 20 minutes.

72b. Hebeloma marginatulum var. proximum
nomen provisiorum

Pileus 1–2.5 cm broad, convex then broadly convex to plane, surface at first whitish canescent, soon glabrous and "Warm Sepia" to "Verona Brown," fading slowly over margin leaving a darker disc, finally concolor overall. Context thin, fragile, odor and taste raphanoid, $FeSO_4$ staining base of stipe dark gray.

Lamellae close, depressed-adnate, broad, finally ventricose, pallid brown when young, near "Sayal Brown" (or darker) at maturity.

Stipe 2–5 cm long, 3–5 mm thick, equal, fragile but not splitting, surface grayish brown to near apex (brunnescent from base upward), apex silky and pallid, no annular zone observed. Veil thin.

Spores 10–12.5 (15) × 6.5–8 μm, inequilateral in profile, ovate in face view, dull clay color in KOH, slowly dextrinoid, apparently smooth as seen in KOH, in Melzer's seen to be minutely punctate.

Hymenium.—Basidia 4-spored, with globules in interior near apex, basal area reddish as seen in sections mounted in Melzer's. Cheilocystidia versiform: 44–100 × 6–11 × 4–5 μm, elongate fusoid-ventricose; and some filamentous, 47–100 × 4–5 μm (apex not enlarged); some basidiolelike with an apical protuberance.

Lamellar and pilear tissues.—Lamellar trama of ± parallel-interwoven broad hyphae red as revived in Melzer's reagent but soon fading to orange-ochraceous; subhymenium of small cells in narrow hyphae. Cuticle of pileus an ixocutis, the hyphae tubular, 1.5–4 μm diam, walls smooth to faintly encrusted. Hypodermium intermediate (cellular and hyphoid intermixed), hyphal walls not conspicuously encrusted; the region reddish brown in KOH and fading on standing. Trama of pileus reddish in Melzer's but quickly fading to ochraceous. Dextrinoid debris present. Clamp connections present.

Habit, habitat, and distribution.—Under spruce, Elk Camp, Pitkin County, Colorado, August 28, 1980, Evenson and Smith 90631 (MICH).

Observations.—In this collection the stipe did not split into segments, the hypodermial elements were only slightly encrusted, and the veil was grayish buff at maturity. The variants of *H. marginatulum* in North America need further study.

73. Hebeloma insigne sp. nov.

Illus. Pl. 7.

Pileus 5–9 cm latus, late convexus, viscidus, obscure vinaceo-brunneus, ad marginem pallide subroseus et appendiculatus; contextus griseus, odor pungens, sapor mitis. Lamellae latae, confertae, brunneomaculatae. Stipes 4–7 cm longus, 1–3 cm crassus, abrupte bulbosus, squamulosus, deorsum tarde brunnescens. Sporae 12–15 × 6.5–8 μm, late inequilaterales, dextrinoideae, verruculosae. Basidia tetraspora, 9–12 μm, clavata. Cheilocystidia 52–70 × 7–9 μm, elongate clavata.

Specimen typicum in Herb. Univ. Mich. conservatum est, Smith 89184; legit prope Elk Camp, Burnt Mt., Pitkin County, Colorado, 29 Aug 1978.

Pileus 5–9 cm broad, broadly convex, becoming plane or nearly so, viscid, opaque, "Pecan Brown" to "Rood's Brown," on the margin "Pale Vinaceous-Pink," margin at first decorated with thin evanescent patches of fibrils from a very scanty veil. Context watery gray fading to whitish, odor pungent, taste mild; FeSO$_4$ slowly staining stipe base grayish; KOH on context no change.

Lamellae broad, close, adnexed, seceding, edges uneven, "Verona Brown," not beaded but edges spotted a darker brown than the faces (from dried droplets ?).

Stipe 4–7 cm long, 1–3 cm thick at apex, equal or nearly so, flared slightly at the base to an abrupt bulb, ± solid but with pith area ± fibrous; becoming brownish within generally; surface with concentric zones of pallid scales, a basal zone present and this easily mistaken for a slight veil, in age glabrescent.

Spore deposit "Warm Sepia" to "Verona Brown" as air-dried (distinctly reddish brown). Spores 12–15 × 6.5–8 μm, warty-rugulose (under a high-dry objective), shape in profile broadly inequilateral, ovate in face view, dextrinoid.

Hymenium.—Basidia 4-spored, 9–12 μm broad near apex, containing refractive globules in many of the cells. Pleurocystidia none. Cheilocystidia 52–70 × 7–9 μm, elongate-clavate, very abundant, hyaline, not agglutinated.

Lamellar and pilear tissues.—Lamellar trama typical for the genus. Cuticle of pileus and ixotrichodermium collapsing in age to an ixolattice, the hyphae 2–4 μm diam and the walls refractive, many with adhering hyaline particles of debris (in KOH mounts). Hypodermium hyphoid, yellowish in KOH fresh, reddish brown in KOH as revived, walls mostly minutely roughened. Tramal hyphae intricately interwoven, the cells greatly inflated, thin-walled, smooth and hyaline. Clamps present.

Habit, habitat, and distribution.—Clustered on a stream bank, in woods of spruce and fir with some pine, near Snowmass Village, at Elk Camp, Pitkin County, Colorado, August 29, 1978 (type, MICH).

Observations.—This is one of the species confused with *H. sinapizans* in North America, and though possessing a "veil," is not closely related to the species of subgenus *Hebeloma*. It is distinguished as a species by the pileal ixotrichodermium, dextrinoid spores, weak $FeSO_4$ stain on the base of the stipe, mild taste, and vinaceous-brown pilei. In Smith 87013 hyaline droplets were present on the gills.

74. Hebeloma kuehneri Bruchet

Bull. Soc. Linn. de Lyon, 39 année: 21. Jun 1970

Illus. Bruchet, l.c., pl. ii.

Pileus 2.5–4 cm broad, obtuse to convex, becoming plano-umbonate or broadly campanulate, subviscid, "Prout's Brown" (dark date brown) beneath a faint hoary coating of whitish fibrils, margin fringed at first, color finally pallid on the margin, the disc about "Wood Brown," opaque at all times. Context watery brown fading to pallid, odor pungent, taste slight; $FeSO_4$ staining base of stipe olive-fuscous.

Lamellae broad, close, adnate, dull pinkish brown ("Pinkish Cinnamon") at maturity, edges serrulate, not beaded and not spotted.

Stipe 2–4 cm long, 3–5 mm thick, equal, hollow, brownish within, surface whitish at first, pruinose above, darkening from the base up; veil remnants very inconspicuous.

Spores 10–14 × 6.5–8 μm, shape in profile ± inequilateral, in face view ovate to broadly fusiform, ± clay color in KOH, slowly but distinctly dextrinoid, surface marbled to faintly rugulose, apex obtuse.

Hymenium.—Basidia 4-spored, 8–11 μm broad near apex. Pleurocystidia none. Cheilocystidia fusoid-ventricose, 34–55 (62) × 7–12 × 3–5 μm, apex obtuse, no secondary septa observed, only one cystidium seen forked at apex.

Lamellar and pilear tissues.—Lamellar trama typical for the genus. Cuticle of pileus a poorly defined ixocutis of hyphae 2–3.5 μm diam, clamped and often with faintly roughened walls, the layer intergrading with the hypodermium. Hypodermium mostly hyphoid, the hyphal walls dark rusty brown and heavily encrusted (as revived in KOH), a redder brown in Melzer's, dextrinoid debris rather abundant in and near the layer. Tramal hyphae typical of the genus and not having red content in mounts in Melzer's. Clamp connections present.

Habit, habitat, and distribution.—Gregarious on duff, Independence Pass area, Pitkin County, Colorado, August 14, 1978, Smith 88982 (MICH).

Observations.—This species is recognized by the very thin pallid veil, broad lamellae and dark brown pileus. It is difficult to recognize in the field. Both alder and conifers were present in the North American habitats where it was found.

75. Hebeloma wells-kemptonae sp. nov.

Pileus 7–13 cm latus, convexus demum late convexus vel planus, subviscidus brunneoincarnatus. Contextus albus, sapor mitis. Lamellae angustae, confertae, pallide brunneae. Stipes 6–9 cm longus, 15–30 mm crassus, deorsum subbulbosus, albidus, siccus, squamulosus. Velum albidum floccosum, evanescens. Sporae 12–17 × 7–9 μm, crassotunicatae, dextrinoideae, inequilaterales. Basidia tetraspora. Pleurocystidia nulla. Cheilocystidia versiformia: cylindrica, fusoid-ventricosa, elongat-clavata, capitato-pedicellata, etc.

Specimen typicum in Herb. Univ. Mich. conservatum est, Wells-Kempton 5304; legit ad viam prope Fairbanks, Alaska, 17 Aug 1971.

Pileus 7–13 cm broad, convex with an inrolled margin, becoming broadly convex to plane, the margin often upraised; color brownish pink becoming browner in age and on drying, tacky (subviscid) when fresh or wet from rains, smooth, appearing slightly appressed-fibrillose. Context white, odor not distinctive, taste pleasant and mild.

Lamellae narrow (4–6 mm), close, adnexed, white becoming pale tan, finally dull brown, edges beaded with hyaline drops when young.

Stipe 6–9 cm long, 15–30 mm thick, equal more or less to a bulbous base, white, dry, scurfy to scaly with scattered floccose fibrillose scales overall (mostly toward the apex), apex often beaded with drops, scales white at first, becoming brown. Context white, stuffed to solid.

Partial veil white, floccose, leaving a floccose band on pileus margin, veil soon evanescent and at times the remnants found near the base of the stipe, pileus margin sulcate from gill impressions in the soft velar tissue.

Spore deposit about "Verona Brown" to "Sayal Brown." Spores 12–17 × 7–9 μm, wall about 0.5 μm thick, surface finally punctate-roughened but long remaining smooth (under high-dry objective), apex having a hyaline spot; color bright clay color to darker in KOH, in Melzer's dark reddish brown (dextrinoid); shape in face view decidedly ovate, in profile inequilateral, apex often snoutlike.

Hymenium.—Basidia large (9–13 μm broad near apex), clavate to cylindric. Pleurocystidia absent. Cheilocystidia clavate to pedicellate-capitate, some more or less utriform, very variable in size: widest cells ± 20 μm broad (± balloon-shaped above a pedicel 5–8 (12) μm diam); clavate-pedicellate cells (28) 33–50 × 9–13 μm; rarely some cells fusoid-ventricose, hyaline at first but all types becoming bunched indiscriminantly and agglutinating, becoming rusty brown both in the wall and context in the process.

Lamellar and pilear tissues.—Lamellar trama typical of the genus except that in Melzer's mounts it becomes red and soon fades to yellowish, the color being located in the cell walls; subhymenium remaining orange in some sections. Cuticle of pileus an ixocutis of slime-coated hyphae 3–6 μm diam, the hyphae widely separated in the layer by the slime. Hypodermium rusty red-brown in Melzer's, ± tawny in KOH, hyphoid, some incrustations present but these not prominent. Tramal body in part red in Melzer's but soon fading (most highly colored near the hypodermium). Clamps present.

Habit, habitat, and distribution.—Collected along a road at University Woods, Fairbanks, Alaska, August 17, 1971 (type, MICH), caespitose. The collectors reported it abundant throughout the area.

Observations.—This veiled species connects to members of subgenus *Denudata* in such characters as the bulkiness of the basidiocarps, the copious formation of droplets on the young gills, the agglutination of the cheilocystidia in age, scarcity of fusoid-ventricose cheilocystidia, the dextrinoid spores and the ± scaly stipe. The versiform cheilocystidia and the ixocutis of the pileus distinguish the species from *H. insigne* which features an ixotrichodermium for a pilear cuticle. The cottony veil would appear to be fairly similar in both. More study of these two in relation to *H. sinapizans* is needed.

76. Hebeloma subrubescens sp. nov.

Illus. Figs. 49–50.

Pileus 2–5 cm latus, convexus demum planus vel late umbonatus, ad marginem albofibrillosus, ad centrum badius ("Prout's Brown" vel "Benzo Brown" demum rufobadius vel rubrovinaceus); odor et gustus mitis demum ± pungens. Lamellae confertae, latae, adnatae, obscure cinnamomeae, serrulatae. Stipes 4–8 cm longus, 3–7 mm crassus, striatus, dissiliens, deorsum demum fulvus. Velum fibrillosum, album demum leviter ochraceum. Sporae 11–14 × 6.5–8 μm, in "KOH" ± leves, non dextrinoideae, inequilaterales. Cheilocystidia clavata, 10–16 μm crassa, vel fusoide ventricosa, 36–72 × 7–16 μm; vel filamentosa, 4–6 μm diam.

Specimen typicum in Herb. Univ. Mich. conservatum est, Smith 90090; legit prope High Alpine Springs, Snowmass Village, Pitkin County, Colorado, ± 11,000 ft elevation, 30 Aug 1979.

Pileus 2–5 cm broad, margin inrolled at first, many with white patches of veil fibrils, disc bay to dark red but at first "Prout's Brown" to "Benzo Brown," becoming redder and paler as the pileus ages, in late button stages the margin at times fringed with fibrils; KOH on cuticle staining it dark brown, with $FeSO_4$ olive-gray (the same on the base of the stipe); odor and taste mild to slightly pungent.

Lamellae broad, close, adnate, dull cinnamon near maturity, edges serrulate, not beaded.

Stipe 4–8 cm long, 3–7 mm thick, equal, twisted-striate, readily splitting lengthwise, slowly becoming rusty brown from the base up, pruinose at the apex; veil white but remains yellowish where the stipe has discolored, only rarely leaving an annular zone where it breaks.

Spores 11–14 × 6.5–8 μm, smooth or in Melzer's faintly punctate, nondextrinoid to weakly dextrinoid (± reddish tawny slowly); shape in face view ovate, in profile inequilateral, apex obtuse, color in KOH pale tawny.

Hymenium.—Basidia 4-spored, 9–12 μm broad near apex, clavate, often with hyaline droplets (in KOH). Pleurocystidia none. Cheilocystidia variable: (1) clavate and 10–16 μm wide; (2) fusoid-ventricose and 36–72 × 7–16 × 4–7 (8) μm obtuse, not agglutinating, wall ± 0.3 μm thick, smooth, some with pale brownish content as revived in KOH; (3) some filamentous, 4–6 μm diam and up to 100 μm long but often with a septum near the basal septum.

Lamellar and pilear tissues.—Gill trama typical for the genus. Cuticle of pileus a poorly developed ixocutis of appressed tubular hyphae 2–5 μm diam and the layer 2–3 hyphae deep. Hypodermium intermediate as to type ("cellular" in tangential sections), cell walls heavily encrusted with patches of dark brown pigment as revived. Clamps present.

Habit, habitat, and distribution.—Scattered on humus under spruce, High Alpine Springs, ± 11,000 ft elevation, above Snowmass Village, Pitkin County, Colorado, August 30, 1979 (type, MICH).

Observations.—The important taxonomic characters of this species are: (1) the poorly developed ixocutis of the pileus; (2) the wide cheilocystidia of various shapes; (3) large inequilateral spores; (4) pattern of color change of the pileus; (5) the readily splitting stipe; and (6) the white veil which discolors where the stipe is discoloring.

77. Hebeloma pallescens sp. nov.

Pileus 1.5–4 cm latus, demum late convexus, saepe ad marginem undulatus, viscidus, glaber, cinnamomeo-brunneus, ad marginem demum griseobrunneus. Contextus pallidus, odor pungens, gustus mitis. Lamellae avellaneae demum argillaceae, latae, confertae. Stipes 2.5–4 cm longus, 4–6 (9) mm crassus, sericeus, deorsum brunnescens. Velum subargillaceum. Sporae 10–13 × 6–7.2 μm, inequilaterales, nondextrinoideae. Basidia tetraspora, 8–10 μm lata. Cheilocystidia 36–54 × 5–6 × 10–15 μm, subventricosa.

Specimen typicum in Herb. Univ. Mich. conservatum est, Smith 89008; legit prope Snowmass Village, Pitkin County, Colorado, 16 Aug 1978.

Pileus 1.5–4 cm broad, convex expanding to plane or the margin decurved, margin finally spreading or uplifted, sometimes wavy, color "Prout's Brown" to "Cinnamon-Brown," gradually paler, the margin brownish gray in age, fading overall to ± alutaceous, subviscid, soon glabrous. Context pallid when faded, odor pungent, taste mild, $FeSO_4$ instantly staining the base of the stipe blackish.

Lamellae broad, close, adnate, avellaneous when young ± clay color in age, not beaded and not staining.

Stipe 2.5–4 cm long, 4–6 (9) mm thick at apex (flared at times), silky, becoming dark brown from base upward; veil material cinnamon-buff, distributed in small patches on lower half; silky near apex (not pruinose), upper half usually pallid.

Spore deposit dull clay color. Spores 10–13 × 6–7.2 μm, dull ochraceous in KOH, pale reddish tan in Melzer's; shape in profile inequilateral, ovate to boat-shaped in face view, surface faintly marbled (under a high-dry objective); apex ± snoutlike on the larger spores, wall not truly dextrinoid.

Hymenium.—Basidia 4-spored, 8–10 μm broad, projecting 10–15 μm when sporulating. Pleurocystidia none. Cheilocystidia 36–54 × 5–6 × 10–15 μm slightly ventricose near the base, apex obtuse, agglutinated in age.

Lamellar and pilear tissues.—Lamellar trama typical for the genus. Cuticle of pileus a poorly developed ixocutis of hyaline appressed refractive hyphae 2.5–3.5 μm diam, clamps present. Hypodermium cellular to hyphoid, tawny in KOH and reddish tawny in Melzer's, dextrinoid debris present in places in the hypodermium, the pilear and the lamellar trama. Tramal hyphae not red as revived in Melzer's.

Habit, habitat, and distribution.—Gregarious under spruce and fir, above Snowmass Village, Pitkin County, Colorado, August 16, 1978 (type, MICH).

Observations.—For a comparison with *H. corrugatum* see that species. *H. pallescens* shows an exceptional amount of dextrinoid debris in its tissues but the spores are not distinctly dextrinoid. It appears to be close to *H. oregonense* var. *atrobrunneum*.

Stirps OREGONENSE

Spores not dextrinoid.

KEY TO SPECIES

1. Taste and usually the odor of the crushed context raphanoid 2
1. Not as above (odor ± pungent in some and in others the taste
   farinaceous to bitter or at least not raphanoid) 9
   2. Spores 9–12 μm long 3
   2. Spores (10) 12–15 μm or more long 5

78. Hebeloma fastibile (Pers. : Fr.) Kummer

Der Führer in die Pilzkunde, p. 80. 1871

Agaricus fastibilis Fr., Syst. Mycol. 1: 249. 1821.

Illus. Lange, Fl. Agar. Dan., pl. 118, fig. f.

Pileus 3–6 cm broad, convex or nearly plane, margin at first inrolled, in age the margin often wavy, glabrous, viscid over the disc, the margin at first decorated with fibrils or patches from the broken veil, color of the disc ± reddish tawny to alutaceous, toward margin paler to pallid buff. Context white, odor and taste raphanoid or taste finally bitterish.

Lamellae sinuate, adnexed or emarginate, pallid at first, slowly becoming cinnamon, edges white fimbriate, broad at maturity, close to subdistant, at times beaded with hyaline drops of liquid.

Stipe 4–6 cm long, 5–10 mm thick, white, firm, fibrillose, equal above the bulbous base, solid or slightly hollow finally, not brunnescent in age; veil cortinate, thin, leaving remnants on upper portion of stipe but seldom forming more than a slight annular zone.

Spores (9) 10–12 × 5.5–6.5 μm, somewhat inequilateral in profile, elliptic in face view, varying to ± boat-shaped (some obscurely snoutlike at apex), ochraceous in KOH, slowly slightly argillaceous in Melzer's (not dextrinoid), surface ± smooth under high-dry objective.

Hymenium.—Basidia 4-spored. Pleurocystidia none. Cheilocystidia 36–76 × 5–12 × 4–6 μm, fusoid-ventricose, filamentous-subcapitate or narrowly clavate, becoming agglutinated in age.

Lamellar and pilear tissues.—Lamellar trama typical for the genus. Cuticle of pileus an ixocutis. Hypodermium hyphoid. Clamps present.

Habit, habitat, and distribution.—On humus in conifer forests, Idaho and Wyoming (?), summer and fall.

Observations.—We have not collected var. *fastibile* in the western area. Kauffman collected it in Boundary County, Idaho, and Simon Davis found it in Wyoming, but we are inclined to question the identity of the Wyoming collection. However, since *H. fastibile* is the type species of the genus we have included it here. The characters emphasized by Moser (1978) are: the stipe is not scaly (but Fries described it as "squamuloso alba"), it is 10–15 mm thick, the pileus is reddish brown on the disc and paler toward the margin, the lamellae are cocoa colored (reddish cinnamon), the odor and taste are radishlike, and the spores are 9–11 × 6–7 μm.

At present we lack an adequate North American collection which has been carefully described in detail and for which we know the $FeSO_4$ reactions (especially on the base of the stipe), the iodine reaction of the spores and of the tramal tissues, the degree of roughness of the spores, the color of the spore print, degree to which the stipe darkens in age (or does not darken), and the color of the veil in unexpanded basidiocarps. The shape of the spores in profile view allows the species to be arranged in either section. Horak's (1968) drawings indicate the spores are distinctly roughened.

Description from Hesler and Smith manuscript

Pileus 3–6 cm broad, convex or nearly plane, yellowish ochraceous or alutaceous whitish or slightly tawny, wavy, obtuse, glabrous, viscid, margin even, pubescent and incurved. Context white; odor raphanoid, taste bitter.

Lamellae broad, subdistant, sinuate to adnexed or emarginate, unequal, pallid, finally becoming cinnamon, edges white-fimbriate, at times beaded with watery drops.

Stipe 4–6 cm long, 5–10 mm thick, white, firm, fibrillose, equal above the bulbous base, solid or slightly hollowed. Veil webby, leaving remnants on upper portion of stipe; sometimes forming a ring.

Spores (9) 10–12 × 5.5–6.5 μm, slightly inequilateral in profile, elliptic in face view, minutely rugulose, with a slight "nose" in some, pale yellowish brown in KOH. Pleurocystidia none. Cheilocystidia 36–78 × 5–9 (12) μm, ampullaceous with a neck, noncapitate or slightly so, some of them narrowly clavate, agglutinating finally.

Pileus cuticle an ixocutis. Pileus trama of radially disposed hyphae. Hypodermium hyphoid. Stipe cuticle of dry, repent hyphae. Caulocystidia occurring in scattered tufts and similar to the cheilocystidia.

Observations.—Although *H. fastibile* occurs in North America, it has been infrequently reported (see Peck [1910] and Kauffman [1918]). It was described by Persoon, and collections are on deposit at Leiden. These have been studied microscopically by both Singer (1961) and Horak (1968). Singer states that the Persoon and the Fries descriptions agree, and that at Leiden there are two Persoonian collections: Nos. L 910. 258-951 and L. 910. 258-593. In the absence of a designated holotype, Singer has selected the latter of the two collections to serve as the lectotype. Despite slight differences in spore size of the two collections, he is of the opinion that both are eligible to serve as the lectotype. The label on the one he selected bears the notation "cum cortina." It was found by Singer to have spores 8.5–10 × 4.5–5.7 μm, melleous, practically smooth, and ellipsoid to fusoid-ellipsoid. He further stated that he was unable to find cheilocystidia (later observed by Horak). In the other Persoon collection, the spores measure 10.2–12 × 5.8–7 μm.

Horak (1968) in his study of the lectotype, gives a full description not only of the microscopic characters observed, but also of macroscopic features. He found spores to be 8–10.5 × 4.5–6 μm, with the wall rugulose, and in shape ellipsoid to almond-shaped. He found cheilocystidia 40–65 × 7–9 μm which were ampullaceous and not capitate or only obscurely so.

Pomerleau (1980) reports *H. fastibile* from Quebec. *H. colora-*

dense is close to *H. fastibile* but differs in the bicolored veil and the pileus is drab to near maturity.

79. Hebeloma latisporum sp. nov.

Pileus (2) 4–8 cm latus, convexus demum planus, glaber, glutinosus, maculatus, pallide argillaceus, odor et sapor raphaninus. Lamellae pallidae demum subfulvae, angustae, adnate confertae. Stipes 5–8 cm longus, circa 1 cm crassus, brunnescens, albofibrillosus. Sporae 9–12 × 6.5–8 μm, in "KOH" subargillaceae, inequilaterales, non dextrinoideae. Basidia tetraspora. Cheilocystidia 36–52 (90) × 4–8 × 8–11 μm, cylindrica vel elongate clavata vel ad basin subventricosa.

Specimen typicum in Herb. Univ. Mich. conservatum est, Smith 55278; legit prope Nordman, Idaho, 22 Oct 1956.

Pileus (2) 4–8 cm broad, convex becoming plane or nearly so, surface glabrous and slimy-viscid, some with watery spots; "Pinkish Buff" overall when fresh, slightly paler in age or when dried. Context thin, odor and taste raphanoid.

Lamellae pallid at first, when mature ochraceous-tawny to "Sayal Brown," narrow, adnate, close, edges not beaded.

Stipe 5–8 cm long, about 1 cm thick, equal or at base slightly enlarged solid, pallid within, slowly becoming dull brown at the base, white-fibrillose and at first with soft patches of veil remnants over the appressed fibrillose layer.

Spores 9–12 × 6.5–8 μm, minutely marbled, ± clay color in KOH, in Melzer's about the same (not dextrinoid), in profile view distinctly inequilateral, ovate in face view.

Hymenium.—Basidia 4-spored, 8–9 μm broad near apex. Pleurocystidia none. Cheilocystidia 36–52 (90) × 4–8 × 8–11 μm, cylindric-clavate to elongate clavate and with a slightly ventricose portion near the base, hyaline, thin-walled.

Lamellar and pilear tissues.—Lamellar trama typical for the genus. Cuticle of pileus a thick ixolattice of hyphae 1.5–2 μm diam, refractive and with clamps. Hypodermium a cellular layer ± hyaline in KOH, beneath it a layer of hyphae yellowish tawny in KOH, encrusting material absent to inconspicuous in this layer. Tramal hyphae typical of the genus but the area beneath the hypoderm reddish in Melzer's and some pigment patches evident. Clamps present.

Habit, habitat, and distribution.—Gregarious in hemlock forests, Priest Lake area, Idaho, October 22, 1956 (type, MICH); also Smith 88817 from below Independence Pass, Pitkin County, Colorado.

Observations.—This is one of the larger veiled species. It differs from *H. fastibile* in its broader spores, the brunnescent stipe and the details of hypodermium.

80. Hebeloma idahoense sp. nov.

Pileus 3–5 cm latus, convexus demum planus vel leviter umbonatus, glutinosus, ad marginem fibrillosus demum glaber, rufobrunneus demum pallidior, odor et gustus raphanoideopungens. Lamellae latae, subdistantes, subcinnamomeae. Stipes 6–11 cm longus, 4–6 mm crassus, anguste clavatus, deorsum brunnescens. Velum pallide ochraceum. Sporae 9–12 × 6–7.5 μm, leves in "KOH," non dextrinoideae. Basidia tetraspora. Cheilocystidia (46) 57–75 × 7–10 × 5–8 μm, ad basin ventricosa.

Specimen typicum in Herb. Univ. Mich. conservatum est, Smith 66049; legit prope Pearl Creek, Valley County, Idaho, 20 Aug 1962.

Pileus 3–5 cm broad, convex becoming plane or with a slight umbo, viscid to slimy, margin decorated with thin patches of grayish to buff-colored veil fibrils, these soon evanescent; surface smooth, shining when ± dried in situ; color "Verona Brown" (a dull reddish brown) where exposed to the sun, margin paler and ± "Pinkish Buff" to "Light Pinkish Cinnamon," opaque at all times. Context whitish, firm, odor and taste ± raphanoid-pungent.

Lamellae broad, subdistant, deeply adnexed, brownish becoming dull cinnamon (± "Sayal Brown") to more cocoa colored, not beaded, not spotted.

Stipe 6–11 cm long, 4–6 mm thick at apex, equal to narrowly clavate (8–12 mm thick at base), whitish at first but soon discolored overall, thinly coated with pallid to buff veil fibrils but no annular zone evident, apex only very weakly scurfy.

Spores 9–12 × 6–7.5 μm appearing smooth in KOH, minutely punctate in Melzer's under oil-immersion lens, not dextrinoid; shape in profile mostly somewhat inequilateral, in face view mostly ovate, apex obtuse (rarely snoutlike).

Hymenium.—Basidia 4-spored, 8–9 μm broad, when sporulating containing droplets as revived in KOH. Pleurocystidia absent. Cheilocystidia abundant, (46) 57–75 × 7–10 × 5–8 μm, elongately fusoid-ventricose to subfilamentous, apex obtuse, walls thin and hyaline, smooth, not agglutinated.

Lamellar and pilear tissues.—Lamellar trama typical of the genus. Cuticle of pileus thick (100 μm or more), in the form of an ixolattice, the hyphae clamped, 1.5–3 (5) μm diam, hyaline in KOH and walls refractive. Hypodermium a thick cellular layer 4–6 cells deep, the walls ± encrusted and the material slowly soluble in KOH, rusty brown when first revived. Pileal context two-layered (duplex) the lower one (near the hymenium) of interwoven hyphae with short inflated cells, the walls smooth in KOH; the upper layer mostly of interwoven narrower more pigmented hyphae but some cells greatly inflated.

Habit, habitat, and distribution.—Scattered under conifers, mostly englemann spruce, Pearl Creek, Valley County, Idaho, August 20, 1962, (type, MICH), and Elk Camp, Burnt Mt., Pitkin County, Colorado, September 1, 1979. Smith 90152 is an additional collection.

Observations.—This species appears to be close to *H. testaceum* sensu Bruchet but the veil is buff colored, the spores are broader, and it is associated with spruce.

81. Hebeloma flaccidum sp. nov.

Pileus 1–3 cm latus, obtusus demum planus vel obscure umbonatus, subviscidus, glaber vel ad marginem sparse fibrillosus; castaneus demum pallidior. Contextus flaccidus, tenuis, odor et sapor raphanicus. Lamellae latae (ventricosae), adnatae, confertae, sordide cinnamomeae. Stipes 3–5 cm longus, 1.5–3 mm crassus, fragilis, dissiliens, brunnescens. Velum fibrillosum, pallidum. Sporae 10–13 (14) × 6–7.5 (8) μm, non dextrinoideae, inequilaterales, leves. Cheilocystidia 50–100 × 4–6 (9) × 3–6 μm, cylindrica vel anguste fusoid-ventricosa, valde elongata.

Specimen typicum in Herb. Univ. Mich. conservatum est, Smith 90452; legit prope Burnt Mt., Pitkin County, Colorado, 1 Aug 1980.

Pileus 1–3 cm broad, obtuse, expanding to plane but often with a slight umbo, surface slightly viscid, glabrous or with thin patches of appressed fibrils along the margin; disc rusty brown to paler ± chestnut brown, marginal zone ± pinkish tan, finally paler overall to nearly white at the margin. Context very thin and lax, odor weakly raphanoid, taste raphanoid; FeSO$_4$ blackening the lower part of the stipe.

Lamellae broad and becoming ventricose, close, adnate becoming subdistant, grayish, finally ± "Sayal Brown," not beaded or spotted.

Stipe 3–5 cm long, 1.5–3 mm thick, fragile, hollow, splitting readily, at first pallid overall, soon brunnescent from base upward. Veil fibrillose, very thin, fibrils pallid.

Spores 10–13 (14) × 6–7.5 (8) μm, pale tawny in KOH, not dextrinoid, inequilateral in profile, in face view ovate, appearing smooth in either KOH or Melzer's reagent.

Hymenium.—Basidia 4-spored, 6–9 μm broad, narrowly clavate, with hyaline globules in interior near apex, not red or orange in Melzer's. Pleurocystidia none. Cheilocystidia various: 50–100 × 4–6 (9) × 3–6 μm, subcylindric and straight to flexuous, or narrowly fusoid-ventricose with greatly elongated necks, apex not or very slightly enlarged, some shaped like hockey sticks, very variable in size from young to old specimens; hyaline in KOH, content homogeneous, walls thin.

Lamellar and pilear tissues.—Lamellar trama of parallel hyphae, the cells mostly elongate and moderately broad (5–12 μm diam); no distinct color reaction in Melzer's. Cuticle of pileus an ixocutis to an ixolattice, hyphae 2–4 μm diam, hyaline, slime not copious. Hypodermium conspicuously cellular in old pilei and with conspicuous scattered incrustations on the cell walls. Tramal body typical of the genus, not colored red as revived in Melzer's (merely yellowish).

Habit, habitat, ad distribution.—Under weeds in a seepage area, pine and spruce nearby and with willow and alder also present, Burnt Mt. area, Pitkin County, Colorado, August 1, 1980 (type, MICH).

Observations.—This species differs from *H. idahoense* in its flaccid context, small size, pallid veil and typically only elongated cells in the hyphae of the pileus context. They have similar odor and taste, spore ornamentation, and cheilocystidia.

82. Hebeloma indecisum sp. nov.

Illus. Figs. 47–48.

Pileus 3–4 cm latus, obtusus vel convexus demum planus, canescens, pallide avellaneus deinde luteobrunneus, ad marginem appendiculate squamulosus. Contextus crassus, albus, odor et gustus subraphanicus. Lamellae pallide avellaneae demum subfulvae, confertae, latae, adnatae. Stipes 4–6 cm longus, 8–12 mm crassus, albidus, deorsum brunnescens, albofibrillosus, saepe annulatus. Sporae 10–13 × 6–7.5 μm, in "KOH" subleves, in Melzer's punctatae, non dextrinoideae. Basidia tetraspora. Cheilocystidia 48–100 × 3–4 × 4–6 μm, elongatoclavata.

Specimen typicum in Herb. Univ. Mich. conservatum est, Smith 89584; legit prope Elk Wallow, Pitkin County, Colorado, 28 Jul 1979.

Pileus 3–4 cm broad (all young), obtuse to convex, becoming ± plane, surface canescent and pale avellaneous at first, about "Buckthorn Brown" to cinnamon-buff when canescence is removed, margin at first decorated with pallid patches of veil fibrils. Context thick, white, odor and taste ± pungent (raphanoid ?), FeSO₄ staining base of stipe olive-black instantly.

Lamellae broad, close, adnate, pale avellaneous when young, becoming dull cinnamon ("Sayal Brown") in age, not beaded.

Stipe 4–6 cm long, 8–12 mm thick, solid, equal, whitish overall but soon becoming bister from the base up, pith remaining pallid; surface heavily white-fibrillose but fibrils slowly discoloring to buff over area where the color of the surface changes, often with a bandlike annulus.

Spores 10–13 × 6–7.5 μm, minutely punctate-roughened in Melzer's, appearing smooth in KOH, in face view ovate to narrowly ovate varying to subfusiform, in profile view ± inequilateral, ± dark clay color in KOH, pale orange-ochraceous in Melzer's (not dextrinoid).

Hymenium.—Basidia 4-spored. Pleurocystidia none. Cheilocystidia 48–100 × 3–4 × 4–6 μm, somewhat agglutinating, filamentous-clavate, flexuous, rarely with a basal enlargement, walls slightly refractive.

Lamellar and pilear tissues.—Gill trama typical of the genus save that in Melzer's a flush of rose occurs which soon fades. It was noted in mounts generally. Cuticle of pileus an ixocutis to an ixolattice, the hyphae 2–3.5 μm diam, many with finely roughened walls (either in KOH or in Melzer's), the walls ± refractive. Hypodermium hyphoid, rusty brown in KOH but no taxonomically significant incrustations were noted, the layer gradually merging with the context, the latter typical for the genus save for an initial flush of rose when mounts are made in Melzer's. Clamp connections generally present.

Habit, habitat, and distribution.—Gregarious on wet soil, Frying-pan River, upstream from Elk Wallow, Pitkin County, Colorado, July 28, 1979, Robert Peabody and A. H. Smith (type, MICH).

Observations.—The inequilateral spores in profile, the thick stipe, the instantly olive-black reaction of FeSO$_4$ on the stipe, the basically yellow-brown pileus when young and fresh, the pungent odor and taste, the brunnescent stipe, and the pallid veil discoloring as the stipe discolors, are the important features.

83. Hebeloma pallido-argillaceum sp. nov.

Pileus 4–5 cm latus, late convexus vel leviter umbonatus, viscidus, ad marginem fibrillosus demum glaber, argillaceus ad marginem demum albidus vel pallidus. Contextus firmus, crassus, pallidus demum albidus, raphanicus. Lamellae latae, confertae, adnatae, sordide cinnamomeae. Stipes 4–7 cm longus, 8–10 mm crassus, aequalis, cavus, brunnescentes, tenuiter fibrillosus. Velum albidum. Sporae 10–13 (14) × 7–8.5 μm, inequilaterales, non dextrinoideae. Cheilocystidia 33–60 × 5–7 × 6–11 μm, elongate clavata rare fusoide ventricosa.

Specimen typicum in Herb. Univ. Mich. conservatum est, Smith 90600; legit prope Elk Camp, Burnt Mt., Pitkin County, Colorado, 28 Aug 1980.

Pileus 4–5 cm broad, obtuse to convex, when expanded often with a slight umbo, at first with faint patches or a fringe of veil fibrils along the margin, soon entirely glabrous, viscid when moist, soon dry, ± pale clay color over the disc and the margin pinkish buff to whitish. Context firm, thick, watery pallid fading to white, odor and taste strongly of radish, $FeSO_4$ staining base of stipe blackish (olive-gray at first).

Lamellae broad, adnate, close, dull "Sayal Brown," not beaded and not spotted.

Stipe 4–7 cm long, 8–10 mm thick, equal, hollow, brunnescent from the base up; surface long remaining whitish, thinly fibrillose from pallid fibrils over lower portion.

Spores 10–13 (14) × 7–8.5 μm, inequilateral in profile, ovate (to ± snoutlike at apex) in face view, punctate finely in Melzer's reagent, not dextrinoid, in KOH dull tawny.

Hymenium.—Basidia 4-spored, 8–9 μm broad near apex, with hyaline globules as revived in KOH, red at first in Melzer's reagent. Pleurocystidia none. Cheilocystidia flexuous-filamentous becoming elongate-clavate, 33–60 × 5–7 × 6–11 μm, or not common but ± fusoid-ventricose with obtuse apex and 34–55 × 7–10 × 5–6 μm, hyaline in KOH.

Lamellar and pilear tissues.—Lamellar trama red in Melzer's in places when first revived; consisting of broad, long and short hyphal cells ± parallel to (finally) interwoven. Cuticle of pileus an ixocutis of hyphae 2–5 μm diam, slime not copious, hyphae tubular and ± weakly ochraceous to hyaline in KOH. Hypoderm hyphoid and poorly differentiated, merely brownish in KOH and incrustations not significant. Trama of pileus of ± interwoven hyphae, red to orange-ochraceous in Melzer's, large inflated cells variously distributed. Clamp connections present.

Habit, habitat, and distribution.—Scattered under spruce and fir, Elk Camp, Burnt Mt., Pitkin County, Colorado, August 28, 1980 (type, MICH).

Observations.—This species is distinct from *H. pseudofastibile* in having most cheilocystidia distinctly enlarged at the apex at maturity and in having broader spores with the apex tending more to snoutlike. The veil fibrils were not observed to become ochraceous, and the stipe was not noticeably fragile.

84. Hebeloma stanleyense sp. nov.

Pileus 1.5–3 cm latus, demum late convexus, ad marginem leviter fibrillosus, triste rufobrunneus demum vinaceobrunneus ("Army Brown"); odor et gustus valde raphanoideus. Lamellae albidae demum pallide rufobrunneae, confertae, demum latae et ventricosae. Stipes 2–4 cm longus, 2–2.5 mm crassus, pallidus. Velum pallidum, fibrillosum. Sporae 10–14 (15) × 6.5–8 μm, limoniformes, rugulosae, non dextrinoideae. Cheilocystidia 32–46 × 7–10 μm, fusoide ventricosa.

Specimen typicum in Herb. Univ. Mich. conservatum est, Smith 46221; legit prope Redfish Lake, Stanley, Idaho, 17 Aug 1954, Bigelow and Smith.

Pileus 1.5–3 cm broad, obtuse to convex, the margin incurved, expanding to broadly convex to plane, margin pallid from a thin coating of veil fibrils, ground color about "Warm Sepia" or a darker red-brown, drying to a pale to medium "Army Brown" or "Verona Brown." Context thin, brownish, odor and taste strongly raphanoid.

Lamellae broad and ventricose, close, adnate, white at first, slowly becoming "Verona Brown" and where colored by spores slightly more rusty cinnamon, not beaded or spotted.

Stipe 2–4 cm long, 2–2.5 mm thick, equal, pallid and thinly fibrillose, gradually becoming dark brown from the base upward, apical area silky; veil scanty, white to pallid.

Spores 10–14 (15) × 6.5–8 μm, distinctly roughened under high-dry objective, rarely slightly calyptrate, pallid singly, dull brownish, about clay color in groups (mounted in KOH), in profile view broadly inequilateral, in face view ovate, not dextrinoid (or at most weakly so in ± 30 minutes).

Hymenium.—Basidia 4-spored, 8–11 μm broad near apex. Pleurocystidia none. Cheilocystidia scattered along the gill edge, subcylindric to fusoid ventricose, 32–46 × 7–10 μm, collapsing in age and often difficult to demonstrate, apex obtuse.

Lamellar and pilear tissues.—Lamellar trama typical for the genus. Cuticle of pileus a thin ixocutis or a "subixocutis"; the hyphae 2–5 μm diam, ± tubular, hyaline, walls somewhat refractive. Hypodermium cellular, of rusty brown elements heavily encrusted, redder brown in Melzer's, no dextrinoid debris observed. Tramal hyphae typical of the genus but in general a red flush present throughout mounts made in Melzer's. Clamp connections present.

Habit, habitat, and distribution.—On duff under pine, Redfish Lake, Stanley, Idaho, August 17, 1954, Bigelow and Smith (type, MICH).

Observations.—The spores readily distinguish this species from *H. mesophaeum* sensu lato, but the pileus does resemble it in color. The critical features for identification are: (1) the large ornamented spores inequilateral in profile view, (2) the very rudimentary ixocutis over the pileus, (3) the heavy incrustations on the walls of the hypodermial elements, (4) the slowly and weakly dextrinoid spores, (5) the association with lodgepole pine. There is a problem with the cheilocystidia: do they collapse as part of their life cycle? On the evidence available, it may be that the ultimate collapse of the cheilocystidia is the endpoint of the agglutination process.

H. stanleyense appears to be close to *H. kuehneri* Bruchet. The American species, however, has a strong raphanoid taste and odor, much smaller cheilocystidia, and *H. kuehneri* apparently favors mossy alpine habitats under willow brush. Its veil is thin and soon obliterated in contrast to that of *H. stanleyense*.

H. clavulipes Romag. is apparently similar to *H. stanleyense* in a number of respects, but differs in its bulbous stipe and finer ornamentation of the spores.

85a. Hebeloma oregonense sp. nov. var. oregonense

Illus. Figs. 43–44.

Pileus 2.5–3.5 cm latus, late convexus demum ± planus, ad centrum glaber, obscure vinaceobrunneus, subviscidus. Contextus mollis, odor et gustus valde raphanoideus. Lamellae latae et ventricosae, subdistantes, demum griseobrunneae. Stipes 2.5–3.5 cm longus, 2–3.5 mm crassus, cavus, pallidus, deorsum valde brunnescens, fibrillosus. Velum pallide ochraceum. Sporae 12–15.5 × 6.5–7.5 μm, limoniformes, ± leves, leviter dextrinoideae. Basidia tetraspora. Cheilocystidia 34–50 (60) × 7–12 μm.

Specimen typicum in Herb. Univ. Mich. conservatum est, Smith 24289; legit prope Mt. Hood, Oregon, 8 Oct 1946.

Pileus 2.5–3.5 cm broad, very broadly convex to nearly plane, disc practically glabrous, margin with silky fibrils from remains of a veil, disc "Warm Sepia" to "Verona Brown" margin near "Cinnamon-Buff" or paler when young, ± canescent overall, very slightly viscid when fresh. Context pallid, thin, soft, odor and taste strongly of radish.

Lamellae very broad and ventricose (5–6 mm), subdistant, depressed-adnate, becoming adnexed, slightly grayer than "Cinnamon-Buff" when young, near avellaneous to "Wood Brown" in age, not beaded.

Stipe 2.5–3.5 cm long, 2–3.5 mm thick, equal, hollow, pallid above, base dingy yellowish and becoming bister or darker from the base upward in age, generally loosely fibrillose from the remnants of the pale buff veil, these fibrils in some specimens arranged in zones over the lower half of the stipe.

Spores 12–15.5 × 6.5–7.5 μm, shape in profile view inequilateral, in face view ovate, appearing smooth under the oil-immersion lens, dingy pallid clay color in KOH, pale reddish tawny in Melzer's (slightly dextrinoid).

Hymenium.—Basidia 4-spored, 23–26 × 7–12 μm, hyaline in KOH. Pleurocystidia none. Cheilocystidia 34–50 (60) × 7–12 × 3–5 μm, fusoid-ventricose, apex obtuse, hyaline, smooth, thin-walled.

Lamellar and pilear tissues.—Lamellar trama typical for the genus. Cuticle of pileus of subgelatinous hyphae 3.5–6 μm diam, hyaline and appressed (forming a "subixocutis"). Hypodermium cellular, the cell walls bister or darker in KOH. Tramal hyphae typical of the genus. Clamps present.

Habit, habitat, and distribution.—Gregarious to scattered on humus in an open area, Frog Lake, Mt. Hood National Forest, Oregon, October 8, 1946 (type, MICH).

Observations.—The narrow, hollow, fragile brunnescent stipe, scarcely viscid pileus, and raphanoid odor and taste, are its major characters in addition to those of the spores.

H. marginatulum Bruchet differs in not having a strongly raphanoid odor and taste. It does have more distinctly ornamented, thick-walled spores, but only a thin cortina. The stipe in *H. oregonense* is fibrillose from the veil and the latter is buff in color. Bruchet illustrates a stipe for *H. marginatulum* which in some basidiocarps is narrowed at the base almost to the degree of producing a short pseudorhiza.

H. oregonense is close to *H. clavulipes* Romag. but has an equal stipe and a colored veil. The two, however, clearly belong in the same stirps.

85b. Hebeloma oregonense var. atrobrunneum var. nov.

Pileus 2–3.5 cm latus, obtusus vel late convexus, canescens demum glaber, viscidus, atrobrunneus. Contextus brunneus demum griseoochraceus, odor pungens, gustus mitis. Lamellae latae, ventricosae, demum subdistantes, pallidae dein "Verona Brown" (rufobrunneae). Stipes 2.5–4 cm longus, 3–6 mm crassus, brunnescens. Velum fibrillosum, griseoochraceum. Sporae 11–14 × 7–8.5 μm, inequilaterales, non dextrinoideae. Basidia tetraspora, 8–10 μm crassa. Cheilocystidia 42–60 × 7–11 × 6–7 μm, cylindrica vel fusoide ventricosa.

Specimen typicum in Herb. Univ. Mich. conservatum est, Smith 89268; legit prope Elk Camp, Burnt Mt., Pitkin County, Colorado, 2 Sep 1978.

Pileus 2–3.5 cm broad, obtuse to convex or at times the margin uplifted, hoary at first, the disc blackish brown beneath the canescence, marginal area slowly becoming dull reddish brown, margin at first faintly decorated with veil fibrils. Context brown fading to grayish buff, odor pungent, taste ± mild, FeSO₄ quickly staining base of stipe olive-black.

Lamellae broad and ventricose, ± subdistant, deeply depressed around the stipe, pallid when young, soon dull reddish brown ("Verona Brown").

Stipe 2.5–4 cm long, 3–6 mm thick, equal, soon a dull dark brown overall beneath the grayish buff veil remnants and at times a thin superior annular zone present when the veil breaks.

Spores 11–14 × 7–8.5 μm, pale clay color in KOH, pale reddish tawny in Melzer's (not truly dextrinoid), faintly marbled; in profile view broadly inequilateral, in face view broadly ovate to broadly subfusiform.

Hymenium.—Basidia 4-spored, 8–10 μm broad. Pleurocystidia none. Cheilocystidia fusoid-ventricose to cylindric and 42–60 × 7–11 × 6–7 μm, hyaline at first, agglutinating into a rusty brown mass in age.

Lamellar and pilear tissues.—Lamellar trama typical of the genus except sections in Melzer's are soon flushed red as is the hymenium also, the color soon fading. Cuticle of pileus a thin ixocutis (easily removed) of narrow hyphae 2–3 μm diam, hyaline, refractive, septa often with a clamp; dextrinoid debris occurring throughout the layer. Hypodermium intermediate as to type, the hyphae heavily encrusted and rusty brown in KOH, redder in Melzer's. Tramal hyphae in Melzer's with red content soon fading to orange-ochraceous; otherwise typical for the genus.

Habit, habitat, and distribution.—Scattered under conifers, Elk Camp, Pitkin County, Colorado, September 2, 1978 (type, MICH).

Observations.—*H. nigellum* Bruchet and *H. oregonense* var. *atrobrunneum* have pileus colors much alike. *H. nigellum,* however, is described as having a raphanoid then bitter taste. Also, in the American taxon, the hyphae of the pileus context are by no means colored like those forming the hypodermium. This we consider an important difference.

86. Hebeloma felleum sp. nov.

Pileus 1–2.5 cm latus, convexus demum expansus, ad marginem subsquamulosus, glabrescens, subviscidus, fulvobrunneus. Sapor amarus; odor pungens. Lamellae latae, confertae, adnatae vel adnexae; subfulvae. Stipes 2–4 cm longus, 2.5–4 mm crassus, aequalis, brunnescens. Velum fibrillosum, pallide argillaceum, sparsim. Sporae (11) 12–14 × 6.5–8 μm, inequilaterales, dextrinoideae. Cheilocystidia 38–67 × 6–12 × 4–6 μm fusoide ventricosa, obtusa; et cylindracea, 24–64 × 4–6 μm, ad apicem obtusa.

Specimen typicum in Herb. Univ. Mich. conservatum est, Smith 90628; legit prope Elk Camp, Burnt Mt., Pitkin County, Colorado, 29 Aug 1980.

Pileus 1–2.5 cm broad, convex becoming broadly convex, at first with inconspicuous patches of pale buff fibrils along the margin, soon glabrous, subviscid and soon dry, evenly colored ± "Cinnamon Brown" but fading on margin first to "Verona Brown" or paler, disc remaining cinnamon-brown. Context white or along the gill-line brownish, odor pungent, taste very bitter; $FeSO_4$ staining base of stipe black.
   Lamellae close, broad, adnate, pinkish buff becoming dull cocoa color ("Sayal Brown").
   Stipe 2–4 cm long, 2.5–4 mm thick, equal, brunnescent from the base upward but at first pallid overall, in age apex typically paler than lower part; fibrillose at first from the buff-colored veil but often glabrous in age (rarely with a thin annular zone of veil fibrils).
   Spores (11) 12–14 × 6.5–8 μm, inequilateral in profile, ovate in face view, smooth in KOH, in Melzer's faintly punctate and weakly dextrinoid, pale tawny as revived in KOH.
   Hymenium.—Basidia narrowly clavate, 4-spored, 8–10 μm broad, often containing fine globules near apex when first revived in KOH, lower portion ± orange-ochraceous in Melzer's. Pleurocystidia none. Cheilocystidia abundant, scattered to bunched, 38–67 × 6–12 × 4–6 μm, fusoid ventricose, apex obtuse; and filamentous and 26–64 × 4–6 μm, not enlarged at apex, hyaline or as revived in KOH, the walls faintly brownish after 15 minutes.
   Lamellar and pilear tissues.—Gill trama typical of the genus except for the orange-yellow tone in Melzer's; the cells elongate and walls thin and smooth, dextrinoid debris present but hardly significant. Cuticle of pileus a very thin cutis to an ixocutis, the layer 2–3 hyphae deep, hyaline or nearly so; clamps present. Hypodermium intermediate (elements of both types present), walls heavily encrusted with dark fulvous pigment particles. Trama of pileus yellow-orange in Melzer's (possibly a red flush at first), dextrinoid debris sparse.
   Habit, habitat, and distribution.—Under spruce Elk Camp, Pitkin County, Colorado, August 29, 1980, Evenson and Smith (type, MICH).
   Observations.—This species is clearly distinct by virtue of the colored veil and very bitter taste. As dried the basidiocarps resemble those of small specimens of *H. mesophaeum* but the features of the spores prevent assignment to that group. It is close to *H. marginatulum* var. *fallax,* especially in view of the slightly colored cheilocystidia in KOH, but that variety has a raphanoid odor and taste and narrower cheilocystidia.

87. Hebeloma perfarinaceum sp. nov.

Pileus 1–2.5 cm latus, obtusus demum obtuse umbonatus; glaber, subviscidus, griseobrunneus demum sordide rufobrunneus. Contextus pallidus, odor nullus, sapor valde farinaceus. Lamellae latae, adnatae, pallide avellaneae demum griseobrunneae vel sordide cinnamomeae. Stipes 2.5–4 cm longus, 3–5 mm crassus, aequalis, albus, demum brunneus ad basin. Velum pallidum. Sporae 12–14 × 6.5–7.5 μm, inequilaterales, non dextrinoideae vel tarde subdextrinoideae. Basidia tetraspora, in Melzer's rubra. Cheilocystidia 40–70 (100) × 4–7 μm, cylindrica vel filamentosa, vel anguste ventricosa ad basin, ad apicum obtusa.

Specimen typicum in Herb. Univ. Mich. conservatum est, Smith 90421; legit prope Elk Wallow, Fryingpan River, Pitkin County, Colorado, 29 Jul 1980.

Pileus 1–2.5 cm broad, obtuse becoming expanded umbonate, margin decurved and with a thin coating of veil fibrils; surface glabrous, scarcely hygrophanous, subviscidus, "Wood Brown" (brownish gray) becoming "Verona Brown" (dull reddish brown), opaque when fresh. Context pallid, odor slight, taste distinctly farinaceous, FeSO$_4$ staining base of stipe dark olive.

Lamellae close, broad, adnate, horizontal, pale avellaneous to wood brown and finally dull cinnamon ("Sayal Brown"), not beaded or spotted.

Stipe 2.5–4 cm long, 3–5 mm thick, equal, evenly dull white overall but soon brunnescent in the base; surface pallid from whitish veil, apex at most only faintly pruinose.

Spores 12–14 × 6.5–7.5 μm, appearing smooth under a high-dry objective, inequilateral in profile, ovate in face view, when fresh slowly weakly dextrinoid, after being dried and standing in the herbarium slightly more dextrinoid.

Hymenium.—Basidia 4-spored, 8–10 μm broad, containing hyaline "oil" droplets; hymenium slowly becoming red in Melzer's. Cheilocystidia 40–70 (100) × 4–7 μm, cylindric to filamentous (straight or flexuous), or narrowly fusoid-ventricose, apex not enlarged, rarely branched, hyaline in KOH. Pleurocystidia none.

Lamellar and pilear tissues.—Lamellar trama typical of the genus but in Melzer's slowly becoming flushed red. Cuticle of pileus an ixocutis, the hyphae tubular and 2–5 μm diam, clamped. Hypodermium hyphoid, poorly differentiated, pigment deposits scattered on the hyphal walls. Tramal hyphae typical of the genus but not red in Melzer's; dextrinoid debris present in an irregular pattern.

Habit, habitat, and distribution.—Clustered under spruce above

Elk Wallow, Fryingpan River, Pitkin County, Colorado, July 29, 1980 (type, MICH).

Observations.—This species appears to be close to *H. kuehneri* but is at once distinguished by its very pronounced farinaceous taste. The hymenium becoming red as revived in Melzer's is an additional distinguishing character. This, however, needs to be verified on European material.

88. Hebeloma lutescentipes sp. nov.

Pileus 2–3.5 cm latus, demum convexus vel plano-umbonatus, glaber, viscidus, ad marginem brunneogriseus, demum rufobrunneus; odor et gustus mitis. Lamellae confertae, latae, obscure fulvae. Stipes 5–7 cm longus, 4–5 mm crassus, fibrillosus, pallidus, tactu ochraceus, deorsum tarde ochraceus. Velum pallidum, sparsim. Sporae 11–14.5 × 7–8.5 μm limoniformes, in "KOH" ochraceae, non dextrinoideae. Basidia 8–10 μm lata, tetraspora. Cheilocystidia 47–70 (90) × 7–12 × 5–7 μm, fusoide ventricosa vel cylindrica.

Specimen typicum in Herb. Univ. Mich. conservatum est, Smith 24018; legit prope Salmon River, Mt. Hood, Oregon, 3,800 ft elevation, 3 Oct 1946.

Pileus 2–3.5 cm broad, obtuse to convex becoming plane or plano-umbonate, soon glabrous and when moist possibly viscid, the marginal area about wood brown and the disc darker but gradually becoming reddish brown ("Verona Brown"), and as dried for the herbarium about the color of pilei of *H. mesophaeum*. Context thin, brownish when moist, odor and taste mild.

Lamellae soon adnexed, close, broad, dull tawny, edges neither beaded nor spotted.

Stipe 5–7 cm long, 4–5 mm thick, equal, fibrillose from a thin pallid veil, at first ± pallid overall but becoming dingy yellowish when bruised or from the base upward in aging.

Spores 11–14.5 × 7–8.5 μm, ochraceous in KOH singly, ± clay color in groups, pale ochraceous in Melzer's (nondextrinoid), surface minutely marbled; shape in profile ± inequilateral in face view ovate to boat-shaped.

Hymenium.—Basidia 4-spored, 8–10 μm broad near apex. Pleurocystidia none. Cheilocystidia 47–70 (90) × 7–12 × 5–7 μm, fusoid-ventricose to cylindric, apex obtuse to slightly enlarged, hyaline, not agglutinated.

Lamellar and pilear tissues.—Lamellar trama typical for the genus. Cuticle of pileus an ixotrichodermium to an ixolattice, hyphae clamped, 1.5–5 μm diam, the walls refractive, the layer easily removed in sectioning. Hypodermium a zone of ± tawny reddish brown hyphae (in KOH), i.e., hyphoid; redder in Melzer's and with dextrinoid debris in the layer. Tramal hyphae typical of the genus. Clamps present on all tissues.

Habit, habitat, and distribution.—Scattered on humus at edge of a beaver pond, East Fork, Salmon River, Mt. Hood, Oregon, October 3, 1946 (type, MICH).

Observations.—The critical combination of features for this species in addition to the staining of the stipe is: (1) the spore shape in profile view, (2) lack of both an odor and a taste, (3) the well-developed ixotrichodermium, and (4) the thin veil.

89. Hebeloma testaceum (Batsch: Fr.) Quélet

Pileus 2.5–4 cm broad, plano-umbonate, glabrous, viscid, margin soon spreading; color ± pinkish brown to pinkish tan over the disc and pinkish buff over the margin, neither canescent nor with watery spots. Context thin, white then brownish, fragile; odor "fresh" (as in the air after a thunderstorm), taste mild to faintly acrid; KOH on cuticle of pileus dull brown; $FeSO_4$ on stipe base no reaction.

Lamellae narrow and not becoming ventricose, crowded, adnate, beaded with droplets at first, ± cocoa-brown at maturity, edges even.

Stipe 3–5 cm long, 3–4.5 mm thick, equal, whitish (pallid), fragile, readily splitting lengthwise, scurfy at the apex, naked below, slowly discoloring (no buttons available), ± honey-brown in age.

Spores 9–12 × 5–6.5 μm, faintly mottled (revived in Melzer's), subfusiform to ovate in face view, in profile view distinctly inequilateral, not dextrinoid.

Hymenium.—Basidia 4-spored, clavate, short-clavate to cylindric, as revived in KOH with relatively few hyaline globules. Pleurocystidia, none seen. Cheilocystidia elongate-clavate to slightly ventricose near the base, 53–86 × 2–7 × 8–12 μm (at apex), conspicuous, not observed agglutinating.

Lamellar and pilear tissues.—Lamellar trama typical of the genus. Cuticle of pileus a thick ixolattice of hyphae 1.5–3 μm diam (possibly originating as an ixotrichodermium), hyaline in KOH. Hypodermium a tangle of hyphoid elements with many of the cells ± inflated, the layer pale fulvous as revived in KOH. Tramal hyphae of pileus merely yellowish in Melzer's.

Habit, habitat, and distribution.—Gregarious on wet (swampy) soil, under conifers, above Elk Wallow, Fryingpan River, Pitkin County, Colorado, August 29, 1979, Smith 90056 (MICH).

Observations.—Further observations preferably on expanding buttons are needed to establish the degree of development of the veil.

As illustrated by Lange (1938) it is extremely light. Our observations to date support this. Also, the FeSO₄ reaction on the base of the stipe should be checked on more mature specimens. The cheilocystidia indicate a relationship to members of subgenus *Denudata* in that most of them are greatly elongated and have an enlarged though not capitate apex. The aspect of the fruiting bodies in the field is that of *H. pascuense* but the inequilateral spores in profile view clearly distinguish it. *H. piceicola* has a dingy buff veil and cheilocystidia having relatively unenlarged apices. Lange's (1938) account covers our specimens rather well. As shown, the stipes are essentially equal and the change to brown in the lower portion is not very conspicuous. The variation in the cheilocystidia, however, deserves further study.

90. Hebeloma pitkinense sp. nov.

Pileus 2–4 cm latus, plano-depressus, viscidus, ad marginem fibrillosus, glabrescens, pallide spadiceus vel fuscospadiceus; odor et gustus mitis. Lamellae pallidae demum pallide alutaceae vel sordide cinnamomeae, latae confertae. Stipes 4–7 cm longus, 4–5 mm crassus, brunnescens. Velum fibrillosum, alutaceum. Sporae 9–12 × 6–7.5 μm, punctatae, in "KOH" argillaceae. Basidia tetraspora, ad apicem 9–11 μm lata. Cheilocystidia 33–52 × 7–9 μm, deorsum subventricosa.

Specimen typicum in Herb. Univ. Mich. conservatum est, Smith 86865; legit prope Snowmass Village, Pitkin County, Colorado, 18 Jul 1976.

Pileus 2–4 cm broad, plano-depressed and margin typically arched slightly, surface viscid, at first the marginal area decorated with patches of pallid veil fibrils, glabrescent; color a pale date brown to "Saccardo's Umber," margin virgate in age in one pileus. Context dark butterscotch color when moist, near cinnamon-buff when faded, soft, fragile, thin, odor and taste not distinctive, FeSO₄ quickly dark green on the lower part of the stipe, less so in the pileus, KOH no reaction.

Lamellae broad, close, adnate, equal or in age broadest near pileus margin, pallid then ± pinkish buff and finally dark dull cinnamon, thin, edges even, not beaded or spotted.

Stipe 4–7 cm long, 4–5 mm thick, equal, tubular, dull brown within, darkening from the base upward, surface fibrillose from the veil, ± glabrescent, fibrils brownish; surface of stipe fibrillose-striate, grayish pallid at or near the apex.

Spores 9–12 × 6–7.5 μm, minutely punctate, clay color in KOH, ovate in face view, obscurely inequilateral in profile or ventral line straight and dorsal line (in optical section) convex.

Hymenium.—Basidia 4-spored, 9–11 μm broad near apex. Pleurocystidia none. Cheilocystidia 33–52 × 7–9 μm, slightly ventricose near the base, neck with wavy walls and apex obtuse, thin-walled, smooth.

Lamellar and pilear tissues.—Lamellar trama typical for the genus. Cuticle of pileus a subixocutis (slime quickly soluble in KOH), hyphae 3–5 μm diam, tubular, clamps present. Hypodermium not distinct from pileus context.

Habit, habitat, and distribution.—Gregarious to scattered under herbaceous vegetation (weeds) in a spruce seepage area, Burnt Mt., Pitkin County, Colorado, July 18, 1976 (type, MICH).

Observations.—The date brown pileus, pallid buff veil, the ± inequilateral spores in profile view, and lack of a differentiated hypodermium are a distinctive combination of features.

91. Hebeloma praelatifolium sp. nov.

Pileus 1–2.5 (3) cm latus, late convexus, viscidus, glaber, cinnamomeo-brunneus ("Cinnamon-Brown"), ad marginem pallidior; odor et gustus mitis. Lamellae perlatae et ventricosae, adnatae, distantes vel subdistantes, griseae demum cinnamomeae. Stipes 3–5 cm longus, 1.5–2 mm crassus, flexuosus griseofibrillosus, deorsum demum obscure brunneus. Basidia tetraspora. Sporae 11–14 × 6.5–7.5 μm, non dextrinoideae. Cheilocystidia 37–52 × 7–11 × 4–6 μm, fusoide ventricosa demum ± filamentosa, vel 40–60 μm longa, filamentosa.

Specimen typicum in Herb. Univ. Mich. conservatum est, Smith 89109; legit prope Savage Lakes, Pitkin County, Colorado, 23 Aug 1978.

Pileus 1–2.5 (3) cm broad, broadly convex, becoming plane and in some the margin paler, fading to tawny around the disc and the margin to pinkish buff, opaque when faded, viscid. Context very thin and pallid, odor and taste mild, $FeSO_4$ staining the base of the stipe olive-fuscous.

Lamellae broad and ventricose, adnate, distant to subdistant, grayish at first (± "Avellaneous"), at maturity pale cocoa-brown, at most slightly beaded.

Stipe 3–5 cm long, 1.5–2 mm thick, equal, flexuous, surface coated with pale gray fibrils of a thin veil which soon vanishes; base becoming dark brown and the color progressing upward to the ± pallid apical region.

Spores 11–14 × 6.5–7.5 μm, ± clay color in KOH, pale tawny in Melzer's (nondextrinoid), very faintly marbled (mounted in Melzer's), shape in profile ± inequilateral, in face view ovate to subelliptic, the apex blunt.

Hymenium.—Basidia 4-spored, 9–11 μm broad near apex. Pleurocystidia none. Cheilocystidia abundant, when young many fusoid-ventricose and 37–52 × 7–11 μm, in age mostly filamentous and 4–6 μm wide (or retaining a ventricose portion below), walls slightly refractive, apex subcapitate in many.

Lamellar and pilear tissues.—Lamellar trama typical of the genus. Cuticle of pileus a thin ixocutis, hyphae 3–4.5 μm diam and the walls of some with incrustations, gelatinization very slight. Hypodermium cellular, with copious pigment deposits on the walls. Tramal hyphae typical of the genus. Clamps present.

Habit, habitat, and distribution.—Gregarious on swampy soil under spruce, Savage Lakes area, Pitkin County, Colorado, August 23, 1978 (type, MICH).

Observations.—This species is distinguished by the pale gray veil, the decidedly ventricose ± distant gills which are avellaneous at first, the nondextrinoid spores, and lack of a distinctive odor and/or taste. The problem with the cheilocystidia is whether or not the ventricose cells actually become filamentous or whether as maturity is reached the cells formed later are mostly filamentous. This accounts for changing ratios of the two types from young to old specimens.

92. **Hebeloma cinereostipes** sp. nov.

Pileus 2–3.5 cm latus, demum late convexus vel planus, subviscidus, ad marginem tenuiter fibrillosus, demum glaber, sordide vinaceobrunneus ("Warm Sepia") vel obscure vinaceobrunneus; odor et gustus mitis. Lamellae latae, confertae, obscure cinnamomeae. Stipes 2–4 cm longus, 3–6 mm crassus, griseofibrillosus, sursum sericeus, deorsum brunnescens. Sporae 11–14 × 5.5–7 μm, tarde subdextrinoideae, subfusiformes. Cheilocystidia 34–65 × 7–12 × 4–6 μm, fusoid-ventricosa.

Specimen typicum in Herb. Univ. Mich. conservatum est, Smith 88733; legit prope Elk Wallow, Fryingpan River, Pitkin County, Colorado, 26 Jul 1978.

Pileus 2–3.5 cm broad, obtuse to convex, becoming broadly convex to plane or the margin remaining decurved, slightly viscid, margin thinly coated with grayish fibrils, disc glabrous and "Warm Sepia" to "Vinaceous-Brown," margin gradually glabrescent. Context watery brown when moist, pallid when faded; odor and taste mild; $FeSO_4$ staining base of stipe blackish.

Lamellae ± broad, close, adnate, about "Sayal Brown" when mature, crisped at times, not beaded and not spotted.

Stipe 2–4 cm long, 3–6 mm thick, equal, grayish when young from a coating of veil fibrils which leave a faint subapical zone in some, apex silky and not truly white, base staining brown and the change progressing upward.

Spores 11–14 × 5.5–7 μm, pale dull ochraceous in KOH, pale tawny in Melzer's (not truly dextrinoid) but slightly if mount is allowed to stand ± 30 minutes; surface faintly marbled, in face view subfusiform to ovate, in profile elongate-inequilateral, apex ± obtuse.

Hymenium.—Basidia 4-spored, 8–10 μm broad near apex. Pleurocystidia none. Cheilocystidia 34–65 × 7–12 × 4–6 μm, narrowly fusoid-ventricose, the neck becoming greatly elongated, apex merely obtuse, agglutinating somewhat.

Lamellar and pilear tissues.—Lamellar trama typical for the genus except that it flushed red at first when mounted in Melzer's. Cuticle of pileus an ixolattice (possibly an ixocutis at first), the hyphae 2–3 μm wide and clamped, some dextrinoid debris present in the layer. Hypodermium cellular to intermediate, yellow in KOH fresh, rusty brown as revived, slowly becoming paler, incrustations not prominent. Tramal hyphae flushed red in sections mounted in Melzer's but soon fading.

Habit, habitat, and distribution.—On mulch in conifer forests, above Elk Wallow, Fryingpan River, Pitkin County, Colorado, July 26, 1978 (type, MICH).

Observations.—The yellow spores in KOH mounts, their tendency to be fusiform when seen in face view and the gray veil are the important characters. It is readily distinguished from *H. praelatifolium* on spore characters. Both have a gray veil.

93. **Hebeloma juneauense** sp. nov.

Pileus 3–5 cm latus, convexus demum planus, viscosus, ad marginem albofloccosus, incarnato ochraceus vel tarde griseoochraceus; tarde leviter amarus. Lamellae latae, albidae demum brunneae, margines saepe guttas liquidas gerentes. Stipes 3.5–5 cm longus, 6–8 mm crassus, subalbus, tarde argillaceus. Sporae 9–12 × 5–6 μm. Cheilocystidia clavata, 43–64 × 7–10 μm.

Specimen typicum in Herb. Univ. Mich. conservatum est, Wells-Kempton 5958; legit prope Juneau, Alaska, 8 Sep 1972.

Pileus 3–5 cm broad, convex, the margin inrolled, expanding to plane; color variable: pinkish tan to grayish ochre, sometimes with brownish irregular watery spots over the surface, viscid when moist slimy, smooth, shiny when faded, appressed fibrillose under the slime, margin of buttons with a narrow white cottony roll of soft tissue (this often sulcate from gill impressions, the layer or roll soon evanescent). Context brownish when moist, fading to white, 3–4 mm thick, odor none, taste hardly distinct but becoming faintly bitter.

Lamellae whitish becoming brown, adnexed, broad (± 4 mm), close, edges serrulate, when young beaded with hyaline drops which as they dry leave the gills spotted from trapped spores in the droplet.

Stipe 3.5–5 cm long, 6–8 mm thick, equal, whitish slowly becoming tan, dry white-pruinose at apex, finely appressed fibrillose, lower down glabrescent; hollow; cortex concolorous with surface. Veil present as slight cottony roll of tissue (see above).

Spores 9–12 × 5–6 μm, distinctly roughened (under high-dry objective), pale clay color revived in KOH, about tawny in Melzer's, shape in profile ± elongate-inequilateral, in face view ovate to boat-shaped or ± broadly fusoid.

Hymenium.—Basidia 4-spored, 7–9 μm broad, projecting when sporulating. Pleurocystidia none. Cheilocystidia elongate-clavate, 43–64 × 7–10 μm (at apex), hyaline, broadest when short and narrowest when elongated. Gill trama typical but flushed diffusely with red in Melzer's. Cuticle of pileus a thick ixolattice (possibly an ixotrichodermium originally), hyphae 1–2 μm diam, ± hyaline, clamped, refractive. Hypodermium bay-brown in KOH, bay-red in Melzer's, cellular-hyphoid (intermediate), walls encrusted. Tramal hyphae typical of the genus. Clamps present.

Habit, habitat, and distribution.—Near the airport, Juneau, Alaska, on sandy silty soil under alder, cottonwood, and spruce (mixed) nearby; September 8, 1972 (type, MICH).

Observations.—The diagnostic features of this species are: (1) the fresh pileus is slimy; (2) the white band of narrow cottony soft tissue present on the pilear margin; (3) young gills beaded with droplets; (4) stipe slowly becoming brown; and (5) the hypodermium bay-red in Melzer's. The habitat is one in which a large number of veiled species is to be expected.

APPENDIX
Selected Extralimital
Species

Species come to be known, for example, as "European," "North American," or "Australian" on the basis of the geographical area from which they were first described. This same pattern is continued on a continental basis and we designate as "Canadian" any species described from Canada, or "Mexican" if described from Mexico. It should be apparent to those using the present publication that many of our "western" species (for both the United States and Canada) will be found in the Great Lakes and northeastern regions generally. This was indicated years ago (Smith, 1947) for species of *Mycena*. It also means that "eastern" species not yet found in the western area will eventually be found there. This is brought about by comparable habitats and identical genera of host trees for mycorrhiza-formers—such as *Hebeloma*.

Consequently we considered it desirable to add a category of extralimital species to include most of those brought to light by the present study. Our presentation here is not to be regarded as complete for the areas indicated.

Kauffman (1918) in section "Indusiati" (the veiled species), included *H. velatum* Pk., *H. fastibile* Fr., *H. mesophaeum* Fr., *H. gregarium,* and *H. pascuense.*

The fact that by present count 20 taxa for the general area, as covered by Kauffman, are recognized is significant and indicates that the diversity found in the western flora will be comparable to that in our eastern and southern areas when these have been adequately studied. The treatment by Murrill (1917) and Peck's contributions over the years come nearest to showing an appreciation of the diversity in this subgenus of *Hebeloma* for North America.

In the present contribution the extralimital species are included in the main keys but are designated by a letter within parentheses: "*Hebeloma sterlingii* (o)" for example. In the appendix the species are arranged in alphabetical order and are numbered consecutively.

Extralimital Species

94. Hebeloma affine sp. nov. (a)

Pileus 1.5–3 cm latus, convexus demum late convexus, obscure cinnamomeus ("Sayal Brown"), tenuiter fibrillosus, glabrescens, ad marginem fibrillose appendiculatus; odor et gustus mitis. Lamellae adnatae, secendentes, incarnatogriseae demum obscure cinnamomeae, subdistantes, latae. Stipes 3–5 cm longus, 3–5 mm crassus, apice pallidus, deorsum brunnescens, fibrillosus, siccus. Sporae 11–15.5 × 7–8.5 μm, late ellipsoideae; leviter rugulosae. Cheilocystidia 46–77 × 5–16 μm, fusoide ventricosa.

Specimen typicum in Herb. Univ. Mich. conservatum est, Smith 42918; legit prope Wilderness Park, Emmet County, Michigan, 21 Sep 1953.

Pileus 1.5–3 cm broad, convex becoming broadly convex, viscid, "Sayal Brown" overall but appearing grayish from a thin coating of fibrils, the margin for a time appendiculate from remnants of the veil. Context watery brownish; odor and taste mild.

Lamellae adnate-seceding, avellaneous when young, becoming dull cinnamon, broad, subdistant, edges crenulate, not beaded or spotted.

Stipe 3–5 cm long, 3–5 mm thick, apex pallid, base brownish, fibrillose, sand adhering to base but base not bulbous, interior stuffed solid. Veil thin, grayish, fibrillose.

Spores 11–15.5 × 7–8.5 μm, very slightly inequilateral in profile, broadly elliptic in face view, wall 0.3–0.5 μm thick very minutely rugulose, (usually appearing smooth or nearly so), yellowish brown in KOH, slowly dextrinoid.

Hymenium.—Basidia 34–38 × 9–10 μm, 4-spored. Pleurocystidia none. Cheilocystidia 46–77 × 5–16 μm, fusoid-ventricose.

Lamellar and pilear tissues.—Gill trama of subparallel to slightly interwoven hyphae 4–7 μm broad. Pileal cuticle an ixocutis. Hypodermium a brown zone of interwoven hyphae. Cuticle of stipe a zone of pallid thick-walled hyphae, some epicuticular hyphae of the stipe incrusted. Clamp connections present. Pileus trama of loosely interwoven hyphae.

Habit, habitat, and distribution.—On sandy soil, edge of beach dunes, Wilderness Park area, Emmet County, Michigan, September 21, 1953 (type, MICH).

Observations.—*H. dunense* Corb-Heim sensu Moser (1978) has smaller spores (10–12.5 × 6–7.5 μm). *H. affine* has large smooth (±) ellipsoid spores—somewhat like those of *H. colvini*. The latter lacks a veil, however. *H. affine* could be forming mycorrhiza with bearberry, dune grasses, species of *Salix* or possibly *Populus*.

95. Hebeloma agglutinatum sp. nov. (b)

Pileus 3–5 cm latus, obtusus demum plano-umbonatus, ad marginem fibrillosus, glabrescens, viscidus vel subviscidus, rufobrunneus. Contextus pallide brunneus, odor et gustus subraphaninus. Lamellae latae, confertae, adnatae, incarnato-alutaceae demum sordide cinnamomeae. Stipes 4–8 cm longus, 4–6 (8) mm crassus, sursum pallidus, deorsum ± cinnamomeus. Velum pallidum, evanescens. Sporae in cumulo cinnamomeae, 8.5–11 × 5–6 μm, ellipsoideae vel ovoideae, subleves, non dextrinoideae. Cheilocystidia fusoide ventricosa vel clavata, 32–46 × 8–14 × 5–7 μm vel 26–45 × 9–14 μm, valde agglutinata.

Specimen typicum in Herb. Univ. Mich. conservatum est, Smith 88328; legit sub *Piceae,* prope Dexter, Michigan, 12 Oct 1977.

Pileus 3–5 cm broad, obtuse at first and margin incurved, expanding to plano-umbonate or the margin finally uplifted, margin at first fringed with remnants of the thin veil and a few thin patches of fibrils over the marginal area, soon glabrous overall, surface viscid but not slimy, soon dry and moderately dull; color when young "Verona Brown" over the disc and the margin ± pinkish buff, as dried (in herbarium) dingy "Fawn Color" to "Wood Brown." Context thin, brownish when fresh, odor and taste raphanoid, $FeSO_4$ olive on base of stipe.

Lamellae broad, close, adnate becoming adnexed, about vinaceous-buff when young and "Sayal Brown" as spores mature or gills have dried; edges serrulate, not beaded or stained.

Stipe 4–8 cm long, 4–6 (8) mm thick, equal, pallid over upper half, "Sayal Brown" below but finally staining bister at the base, as dried concolorous with pileus margin above and darker below. Veil pallid, remnants soon vanishing.

Spore deposit ± "Sayal Brown" as air-dried. Spores 8.5–11 × 5–6 μm, ellipsoid to ovoid, smooth (under high-dry objective), in profile often obscurely bean-shaped, very pale (± pinkish buff) mounted in either KOH or Melzer's (not dextrinoid).

Hymenium.—Basidia 4-spored, 8–10 μm broad near apex, projecting prominently when sporulating. Pleurocystidia none. Cheilocystidia small and gill edge distinctly refractive at times in KOH, the cystidia clavate to (mostly) fusoid-ventricose, 26–45 × 9–14 μm if clavate and 32–46 × 8–14 × 5–7 μm if fusoid-ventricose with a short neck and obtuse to subcapitate apex, in age agglutinated into an amorphous mass.

Lamellar and pilear tissues.—Lamellar trama typical of the genus but some dextrinoid debris becoming evident and hymenium and adjacent zone ± ochraceous-orange in Melzer's. Cuticle of pileus a thin but distinct ixocutis, hyphae with refractive walls and walls smooth to finely roughened, 2.5–3.5 μm diam and tubular, clamps present. Hypodermium cellular, walls dark brown in KOH, reddish brown in Melzer's, at most not heavily encrusted, no dextrinoid debris observed in or near the layer. Tramal hyphae typical of the genus. Hyphae generally lacking red content in Melzer's.

Habit, habitat, and distribution.—Gregarious under spruce, Dexter, Michigan, October 12, 1977 (type, MICH).

Observations.—This species fruited in groups of a hundred or more basidiocarps in one area of a spruce plantation in company with a few other closely related species of *Hebeloma*. It is one of the few veiled species of this genus in which the cheilocystidia begin to agglutinate at a relatively early stage of maturity. It differs from *H. mesophaeum* in having smaller cheilocystidia and grayish incarnate gills when young.

96. Hebeloma angustifolium sp. nov. (c)

Pileus 2.5–4 cm latus, demum campanulatus vel plano-umbonatus, valde viscidus, ad marginem fibrillosus, glabrescens, "Cinnamon-Brown" ad centrum, sordide cinnamomeus ad marginem. Contextus fragilis, brunneus, odor pungens, gustus tarde subamarus. Lamellae confertae, angustae, adnatae vel demum subsinuatae, griseobrunneae tarde rufobrunneae, ad marginem brunneomaculatae. Stipes 5–10 cm longus, 2.5–4 mm crassus fragilis, avellaneus, deorsum demum vinaceo-brunneus, dissiliens. Spores 8–10 (11) × 5–6 μm, non dextrinoideae, ellipsoideae vel subovoideae, subleves. Cheilocystidia fusoide ventricosa 32–48 (57) × 7–10 × 3.5–5 μm.

Specimen typicum in Herb. Univ. Mich. conservatum est, Smith 88295; legit prope Chelsea, Michigan, 11 Oct 1977.

Pileus 2.5–4 cm broad, obtuse becoming plano-umbonate to campanulate, margin usually decurved, surface slimy-viscid, decorated with patches of the veil along the margin, disc glabrous; color ± "Cinnamon-Brown" over the disc and the margin "Sayal Brown," veil patches cinnamon-buff; surface opaque at all times. Context watery brown, thin, brittle, odor ± pungent, taste slowly disagreeable; FeSO$_4$ staining base of stipe olive-fuscous; KOH on pileus cuticle no reaction.

Lamellae crowded, narrow, adnexed, seceding, avellaneous to wood brown to "Verona Brown," staining slightly to cocoa-brown, edges even and not beaded.

Stipe 5–10 cm long, 2.5–4 mm thick, strict, brittle, equal, avellaneous above midportion at first, lower down coated with cinnamon-buff fibrils from the veil, soon becoming dark vinaceous-brown from the base up, splitting longitudinally into segments readily when broken.

Spores 8–10 (11) × 5–6 μm, ± pale snuff brown in KOH, paler in Melzer's and minutely marbled, ovoid to ellipsoid, some in profile view obscurely inequilateral, not dextrinoid.

Hymenium.—Basidia 4-spored, 7–9 μm wide near apex. Pleurocystidia none. Cheilocystidia mostly fusoid-ventricose, 32–48 (57) × 7–10 × 3.5–5 μm, apex obtuse, hyaline, scattered.

Lamellar and pilear tissues.—Gill trama typical of the genus, the orange-reddish tone becoming orange-ochraceous on standing (in Melzer's mounts). Cuticle of pileus an ixocutis to an ixolattice, the hyphae with gelatinizing walls and often collapsed, clamps present but difficult to locate. Hypodermium ± cellular in tangential section and tawny-brown in KOH, redder brown in Melzer's, very little encrusting material and no dextrinoid debris seen. Tramal hyphae typical for the genus: ± hyaline in KOH, reddish tan in Melzer's (color in the wall).

Habit, habitat, and distribution.—Under shrubs, near Chelsea, Michigan, October 11, 1977 (type, MICH).

Observations.—The readily splitting and brittle stipe, staining of the gills, slimy-viscid pileus and cinnamon-buff veil are distinctive. It is close to *H. brunneomaculatum,* an Alaskan species, but the latter has a peculiar odor and a mild taste along with a less fragile stipe, to say nothing of the difference in habitat.

97. Hebeloma bicoloratum sp. nov. var.
bicoloratum (d)

Pileus 2–4 cm latus, demum late convexus vel subplanus, saepe obtuse umbonatus, viscidus, at centrum glaber, ad marginem fibrilloso-areolatus; ad centrum "Warm Sepia" (sordide fulvus), tarde ("Sayal Brown") obscure cinnamomeus, tarde glabrescens. Contextus brunneus demum pallidus, fragilis, odor nullus, sapor mitis. Lamellae demum ventricosae, confertae, adnatae, demum obscure cinnamomeae. Stipes 3–7 cm longus, 4–7 mm crassus, pallidus deorsum brunnescens, fibrillosus, saepe subannulatus. Velum bicoloratum: cortina album, extus pallide argillaceum. Sporae 7–9 (10) × 4.5–5.5 μm, in "KOH" argillaceae, non dextrinoideae ovoideae vel ellipsoideae, subleves. Basidia tetraspora. Cheilocystidia 36–57 × 7–12 × 4–5 μm, fusoid-ventricosa, obtusa, rare clavata vel versiformes.

Specimen typicum in Herb. Univ. Mich. conservatum est, Smith 88316; legit sub *Piceae,* prope Dexter, Michigan, 12 Oct 1977.

Pileus 2–4 cm broad, obtuse to convex, the margin incurved, expanding to plane or with a low umbo, margin appendiculate and at times remaining decurved, disc glabrous and viscid, marginal area covered by matted veil remnants, these pallid to dingy buff in color and often obscuring the ground color, disc "Warm Sepia" becoming paler to ± "Sayal Brown." Context brown fading to pallid, rigid and brittle, KOH no reaction, FeSO$_4$ olive-fuscous on base of stipe, odor and taste mild.

Lamellae becoming ventricose, pallid to brownish at first, becoming ± "Sayal Brown," close, adnate, edges pallid and not beaded or stained.

Stipe 3–7 cm long, 4–7 mm thick, equal, tubular, pallid at first, soon "Sayal Brown" to near the apex; surface decorated with veil remnants to the subapical annular ragged zone; outer veil cinnamon-buff, inner veil (cortina) white.

Spores 7–9 (10) × 4.5–5.5 μm, clay color in KOH, pallid in Melzer's (not dextrinoid), the surface obscurely mottled, shape in profile obscurely inequilateral, in face view ovate to elliptic.

Hymenium.—Basidia 4-spored, 7–8 μm broad near apex. Pleurocystidia absent. Cheilocystidia 36–57 × 7–12 × 4–5 μm, fusoid-ventricose, neck becoming elongated, apex obtuse; some clavate and sending out ± eccentric protrusions of various lengths (one to two to a cell).

Lamellar and pilear tissues.—Cuticle of pileus an ixocutis and often with veil hyphae simulating an upper thin layer (stuck to the slime-layer), hyphae of the ixocutis 2–3.5 μm diam, wall gelatinizing (some hyphae remaining collapsed), clamps present. Hypodermium hyphoid to cellular, the layer dark brown in KOH, in Melzer's a redder brown and pigmented deposits evident but no dextrinoid debris. Tramal hyphae typical for the genus, but at first with a red flush in Melzer's but this soon fading.

Habit, habitat, and distribution.—Gregarious under spruce, Dexter, Michigan, October 12, 1977 (type, MICH).

Observations.—This variant occurred by the hundreds in the spruce plantation between October 10 and 15. The heavy bicolored veil and mild taste were observed on hundreds of fruiting bodies.

98. Hebeloma gregarium Peck (e)

Ann. Rep. N.Y. State Mus. 49: 18. 1896

Pileus 2.5–3.5 cm broad, hemispheric to convex to obtuse or with a small inconspicuous umbo, surface slightly viscid when moist, becoming radiately rimose by old age, glabrous except for the slight coating of velar fibrils over the marginal area; color pale ochraceous or with the disc more or less reddish tawny. Context whitish, odor of crushed tissue raphanoid.

Lamellae thin, close, adnate, whitish at first, then dull cinnamon, broad.

Stipe 3–5 cm long, 2–4 mm thick, slender, stuffed becoming hollow, whitish fibrillose, slightly pruinose above, with a faint pallid annular zone of fibrils from the broken veil, base not discoloring appreciably.

Spores (8.5) 9–11 × 4.5–7 μm, in face view elliptic to ovate, in profile slightly inequilateral, ornamentation very faint, wall thin and spores in KOH merely pale yellow, not dextrinoid.

Hymenium.—Basidia 4-spored. Pleurocystidia none. Cheilocystidia 28–54 × 5–9 × 4–6 μm, subcylindric to fusoid-ventricose, apex obtuse to somewhat capitate, hyaline.

Lamellar and pilear tissues.—Lamellar trama of narrow (3–5 μm diam), subparallel hyphae pallid in KOH, cells ± elongate. Cuticle of pileus an ixocutis, the hyphae 2–4 μm diam, hyaline, clamps present. Hypoderm "cellular" (Hesler).

Habit, habitat, and distribution.—Gregarious on sandy soil in heathy places, Delmar, New York, late fall. Peck's type.

Observations.—This species is closely related to *H. kauffmanii, H. subrimosum,* and *H. pascuense.* It differs from *H. kauffmanii* in the tendency for the pileus to become rimose, in having a raphanoid odor, and a bitter to disagreeable taste. This last observation was made by G. E. Morris, November 2, 1908 on a collection identified by Peck. From *H. pascuense* it differs in its rimose pileus, broad gills, and in having a cellular hypodermium. From *H. subrimosum* it differs in having a raphanoid odor, in the stipe not darkening below, and in the tendency of the spores to be ± inequilateral in profile view.

99. Hebeloma hydrocybeoides sp. nov. (f)

Pileus 2–3.5 cm latus, convexus demum late expansus vel planus, viscidus nitens laete fulvus demum alutaceus; sapor amarus; odor mitis. Lamellae confertae latae adnatae demum sinuatae, cinnamomeobrunneae. Stipes 2.5–5 cm longus, 2–4.5 mm crassus, aequalis, cavus, albidus, deorsum tarde brunnescens, leviter fibrillosus; velum sparsum, evanescens. Sporae 7–9.5 × 5–6 μm, valde dextrinoideae, rugulosae, fulvae in "KOH." Cheilocystidia 23–37 × 5–9 μm vel 36–43 × 4.5–6 μm et subfilamentosa. Cuticula pileorum crassa ixocutis est.

Specimen typicum in Herb. Univ. Mich. conservatum est, Smith 36228; legit prope Dixboro, Michigan, 17 Oct 1950.

Pileus 2–3.5 cm broad, convex with the margin inrolled, expanding to plane or nearly so, margin often remaining decurved, surface glabrous, viscid and shining, "Mars Brown" to "Cinnamon-Brown" fading to tan. Context brown, taste bitter, odor mild.

Lamellae close, adnexed, finally moderately broad, cinnamon-brown fresh but paler as dried, edges even and not beaded.

Stipe 2.5–5 cm long, 2–4 mm thick, equal, hollow, whitish, fibrillose, a veil absent (young fresh specimens examined) but the fibrils on the stipe may be indicative of a rudimentary veil; brown in the base and where the fibrillose coating has been removed, merely pallid brownish below in herbarium specimens.

Spores 7–9.5 × 5–6 μm, russet in KOH, strongly dextrinoid, distinctly warty-rugulose under high-dry objective; in profile view inequilateral, ovate in face view.

Hymenium.—Basidia 4-spored, 6.5–7.5 μm wide near apex. Pleurocystidia none. Cheilocystidia fusoid-ventricose to subcylindric, 23–37 × 5–9 μm and fusoid-ventricose, or 36–43 × 4.5–6 × 3–5 μm and ± filamentous, elongating greatly as aging takes place.

Lamellar and pilear tissues.—Lamellar trama hyaline in KOH, typical of the genus. Cuticle of pileus an ixocutis to an ixolattice, thick, hyphae widely spaced, gelatinized in some, 1.5–3 μm diam. Hypodermium hyphoid, the area poorly defined. Tramal hyphae typical of the genus. Clamps present.

Habit, habitat, and distribution.—On soil in a meadow, Dixboro, Michigan, October 17, 1950 (type, MICH).

Observations.—The color of the spores is that of *Cortinarius* as is also the ornamentation. On the fresh material no veil extended from the cap margin to the stipe, but the fibrillose coating of the stipe may indicate the presence of one in very young buttons. The combination of the types of cheilocystidia and pattern of delayed elongation along with the well-developed ixocutis of the pileus indicate that the specimens belong in *Hebeloma*.

100. Hebeloma littenii sp. nov. (g)

Pileus 2–6.5 cm latus, plano-convexus, ad marginem undulatus, hygrophanus, glaber, brunneus, demum sordide aurantio-cinnamomeus; odor et gustus mitis. Lamellae subdecurrentes vel adnatae, confertae, latae, pallide argillaceae demum sordide cinnamomeae. Stipes 3–6 cm longus, 5–13 mm crassus, leviter fibrillosus, deorsum brunnescens. Velum pallidum, sparsum. Sporae 10–12 × 6–7.5 μm, non dextrinoideae, in cumulis "Sepia," ellipsoideae vel ovoideae. Cheilocystidia 32–63 × 5–11 × 4–5 μm, fusoid-ventricosa vel ovata, 9–12 μm lata.

Specimen typicum in Herb. Univ. Mich. conservatum est, legit in Mt. Desert Isl., Hancock County, Maine, 29 Oct 1980; W. Litten.

Pileus 2–6.5 cm broad, plano-convex, becoming plane; margin thin and at times becoming wavy, surface hygrophanous, glabrous, when moist light brown (ISCC-NBS) to light reddish brown (before fading to moderate orange), rarely subrimose (in age irregularly), as dried the pileus evenly ± "Verona Brown." Context thick on the disc, very thin at margin, translucent yellowish brown when fresh, fading to almost white and then opaque; odor and taste not distinctive.

Lamellae decurrent by a tooth when young, later adnate, close, moderately broad; edges eroded; surface light tan when young, becoming darker (brown) in a "patchy pattern" (Litten).

Stipe 3–6 cm long, 5–13 mm thick (mostly about 1 cm thick), stuffed; surface somewhat fibrillose; color pallid at apex, darker brown near base; veil fibrillose, pallid, all traces soon vanishing.

Spore deposit as air dried ± "Sepia." Spores 10–12 × 6–7.5 μm, thin-walled, minutely punctate under an oil-immersion objective, nearly hyaline in KOH, not dextrinoid; in face view elliptic, in profile elliptic to oblong.

Hymenium.—Basidia 4-spored, as revived in KOH with numerous hyaline globules, 7–9 μm broad when sporulating. When first revived in Melzer's dull red but soon fading. Pleurocystidia none. Cheilocystidia abundant, 32–63 × 5–11 × 4–5 μm, fusoid-ventricose, some ovoid and 9–12 μm diam; becoming considerably elongated in age and apex in some subcapitate; content often dull ochraceous but homogeneous in KOH as revived.

Lamellar and pilear tissues.—Cuticle of pileus a thin ixocutis or a "subixocutis" (only a few hyphae deep and these scarcely gelatinous), hyphae 2–3 (4) μm diam, tubular, clamps present. Hypodermium of inflated but elongated hyphal cells, rusty brown as revived in KOH and distinctly encrusted (the layer intermediate between hyphoid and cellular). Tramal hyphae of pileus and lamellae typical of the genus but red as revived in Melzer's, soon fading.

Habit, habitat, and distribution.—Gregarious to caespitose on a lawn under *Pinus strobus,* following several weeks of killing frost at night, Bar Harbor, Mt. Desert Isl., Hancock County, Maine, October 29, 1980 (type, MICH).

Observations.—A thin fibrillose veil is still visible on the stipe in young fruiting bodies, but the pileus margin lacks the patches of fibrils characteristic of *H. mesophaeum. H. littenii* appears to be closest to *H. subrimosum* but is not *Inocybe*-like in the appearance of the fruiting bodies, and the western species does not have a translucent cap when moist. The dark-colored spore deposit of *H. littenii* is in contrast to its nearly hyaline spores as mounted in KOH and viewed under the microscope.

101. Hebeloma mesophaeum var. longipes var. nov. (h)

Pileus 2–6 cm latus, campanulatus vel plano-umbonatus, viscidus, argillaceus, ad marginem fibrillose appendiculatus, glabrescens. Contextus pallide argillaceus; odor et gustus raphanoideus. Lamellae pallidae dein avellaneae demum sordide cinnamomeae. Stipes 5–11 cm longus, 5–10 mm crassus, sursum pallidus, deorsum brunnescens, sparse fibrillosus. Velum pallidum, sparsum, fibrillosum, evanescens. Sporae 8–10 × 5–6 μm, ovoideae vel ellipsoideae, non dextrinoideae, subleves, in "KOH" pallide argillaceae. Cheilocystidia 32–41 (60+) μm longa, 6–9 μm lata ad basim, sursum 4–6 μm lata.

Specimen typicum in Herb. Univ. Mich. conservatum est, Smith 64681; legit sub *Pini,* prope Dexter, Michigan, 14 Oct 1961.

Pileus 2–6 cm broad, obtuse to campanulate, expanding to broadly campanulate, sometimes plane or with a low umbo, surface viscid, margin fringed with fibrils but soon glabrescent, disc clay color to cinnamon-buff, margin pinkish buff or paler, (as dried in the herbarium) the disc alutaceous and the margin dull pinkish buff. Context thin, pale dingy buff when faded, odor and taste raphanoid, KOH and FeSO₄ neither one staining the cuticle or context.

Lamellae whitish becoming avellaneous and finally pale "Sayal Brown," close, broad, adnate to adnexed, edges slightly uneven, not beaded or spotted.

Stipe 5–11 cm long, 5–10 mm thick at apex, equal or nearly so, whitish and silky at apex, the fibrillose zone left by the veil evanescent, thinly fibrillose below the veil-line, color in base becoming dark yellow-brown and the change progressing upward. Veil pallid in button stages, thin and not discoloring on the stipe.

Spores 8–10 × 5–6 μm, ovoid to ellipsoid, not dextrinoid, apparently smooth, pale buff in KOH, wall thin.

Hymenium.—Basidia 4-spored, 7–8 μm broad near apex. Pleurocystidia none. Cheilocystidia 32–41 (60+) × 6–9 × 4–6 μm, fusoidventricose but in age with the neck greatly elongated; some filamentous and 40–55 × 5–6 μm, apex obtuse to subcapitate.

Lamellar and pilear tissues.—Lamellar trama typical for the genus. Cuticle of pileus a well-defined ixocutis of hyphae 1.5–3 μm diam, clamped, walls refractive in KOH. Hypodermium hyphoid, scarcely any difference between it and the tramal tissue proper. Tramal hyphae typical of the genus.

Habit, habitat, and distribution.—Gregarious under white pine, Dexter, Michigan, October 14, 1961 (type, MICH).

Observations.—This variety is distinct by virtue of the clay-colored pileus, hyphoid hypodermium, long stipe and thin veil.

102. Hebeloma naucorioides sp. nov. (i)

Illus. Pl. 8.

Pileus 1.5–3.5 cm latus, convexus demum obtuse umbonatus vel subplanus, tenuiter fibrillosus, glabrescens, subviscidus, incarnatogriseus vel avellaneus. Contextus mollis; odor fragrans; sapor mitis. Lamellae latae, confertae incarnato-brunneae ("Fawn Color"). Stipes 4–7 cm longus, 2.5–3.5 mm crassus, cartilagineus, cavus, "Fawn Color," deorsum obscurior, griseo-fibrillosus, glabrescens. Sporae 9–12 × 5.5–6.5 (7) μm, ellipticae vel subellipticae, in laterales subellipticae, non dextrinoideae. Cheilocystidia 37–63 × 7–11 × 4–5 (7–8) μm, rare subcapitata.

Specimen typicum in Herb. Univ. Mich. conservatum est, Smith 32-580; legit prope Lakeland, Livingston County, Michigan, 15 Oct 1932.

Pileus 1.5–3.5 cm broad, very nearly convex to obtusely umbonate at first, becoming convex or nearly plane, margin long remaining incurved; surface at first covered by sparse appressed fibrils from the veil, these often aggregated into fascicles near the margin, glabrescent, subviscid to viscid, margin even; color "Vinaceous-Fawn" on the disc to more brownish avellaneous near the margin. Context very soft and fragile, odor faintly but distinctly fragrant, taste mild.

Lamellae broad, close, rounded-sinuate, "Fawn Color" becoming "Sayal Brown," edges even, neither beaded nor spotted.

Stipe 4–7 cm long, 2.5–3.5 mm thick, equal, cartilaginous, hollow, "Fawn Color" but becoming darker below; grayish fibrillose from the remains of the grayish fibrillose veil.

Spores 9–12 × 5.5–6.5 (7) μm, elliptic to obscurely ovate in face view, in profile subelliptic to very weakly inequilateral in profile, apex rounded, wall ± buffy hyaline in KOH, not dextrinoid, clay color in Melzer's and very minutely punctate.

Hymenium.—Basidia 4-spored, 8–9 μm broad near apex, not red in Melzer's (but material poorly dried). Pleurocystidia none. Cheilocystidia 37–63 × 7–11 × 4–5 (7–8) μm, elongate fusoid-ventricose finally but many not greatly elongated, some in age spathulate at the apex (with a tendency to fork), often shaped like a hockey stick in age.

Lamellar and pilear tissues.—Lamellar trama typical of the genus, not truly red in Melzer's but in places orange-ochraceous. Cuticle of pileus an ixotrichodermium becoming an ixolattice, hyphal ends 1.5–2 μm diam, the elements tubular, not infrequently branched, copious slime in the layer in material revived in KOH; basal to the ixotrichodermium are found ± hyaline hypodermial (?) hyphae 5–12 μm diam with hyaline ± refractive walls in KOH. Hypodermium proper a narrow zone of fulvous (in KOH) hyphae with short to elongated cells 8–18 μm diam. Tramal hyphae yellow in Melzer's. These seemingly typical of the genus.

Habit, habitat, and distribution.—Gregarious in muck on swampy ground, Lakeland, Livingston County, Michigan, October 15, 1932 (type, MICH).

Observations.—The fragrant odor, *H. mesophaeum*-like spores, pale gray veil, and pale vinaceous-brown ("Fawn Color") pileus are an important combination of characters in this genus.

103. Hebeloma pascuense Peck (j)

Ann. Rep. N.Y. State Mus. 53: 844. 1900

Illus. Peck, l.c., pl. C, bottom.

Pileus 2–5 cm broad, convex becoming nearly plane, viscid when moist, obscurely innately fibrillose, brownish clay color, often darker or rufescent on the disc, margin at first slightly whitened by the thin webby veil. Context whitish, taste mild, odor weak, resembling that of radishes.

Lamellae close, rounded behind, adnexed narrow to moderately broad (see Peck's illustration), whitish becoming pale ochraceous to more orange-cinnamon.

Stipe short, 3–5 cm long, 3–5 mm thick, firm, equal, solid, near apex ± pruinose ("mealy"), whitish or pallid, apparently not darkening.

Spores 8.5–10.5 × 6–6.5 μm nearly hyaline in KOH and the same in Melzer's (not dextrinoid); thin-walled, in face view ovate, in profile ± ovate to obscurely inequilateral apex rounded to obtuse, surface practically smooth (under high-dry lens).

Hymenium.—Basidia 4-spored, clavate, 7–9 μm wide near apex. Pleurocystidia none. Cheilocystidia 40–65 × 6–10 × 3–4 × 3–5 μm, elongate fusoid-ventricose, apex weakly enlarged, neck flexuous, hyaline in KOH, not agglutinated.

Lamellar and pilear tissues.—Lamellar trama typical of the genus, cells of hyphae only weakly inflated. Cuticle of pileus an ixocutis, hyphae 2–5 mm wide, colorless. Hypodermium hyphoid, colorless in KOH, pale orange-brown in Melzer's; amyloid debris present in the mounts. Tramal hyphae of the pileus typical for the genus; clamps present.

Habit, habitat, and distribution.—This species has not yet been established as occurring in western North America, and has been confused with a number of species in the central and eastern states. It is found growing gregariously in old fields featuring poor soil. Our description is based on a study of the type and Peck's original account in order to avoid further confusion.

Observations.—The dried specimens remind one of those of *H. mesophaeum* but it is clear from Peck's painting that the stipe does not darken appreciably even at the base. At present we regard the amyloid debris mentioned above as an artifact.

104. Hebeloma proximum sp. nov. (k)

Pileus 1.5–3 (5) cm latus, obtusus, demum plano-umbonatus, viscidus, pallide incarnato-cinnamomeus, ad centrum rufobrunneus; odor et gustus valde raphaninus. Lamellae pallide incarnato-cinnamomeae demum subfulvae, latae, confertae, adnatae. Stipes 6–10 cm longus, 4–9 mm crassus, solidus, dissiliens, pallide cinnamomeus. Velum fibrillosum, sparsum, pallidum. Sporae 8–10 × 5–5.5 μm, non dextrinoideae, ellipsoideae vel ovoideae, subleves. Basidia tetraspora. Cheilocystidia fusoid-ventricosa, 33–47 × 7–12 μm vel cylindrica et ± 70 × 5–6 μm, ad apicem obtusa.

Specimen typicum in Herb. Univ. Mich. conservatum est, Smith 75218; legit sub *Piceae,* prope Hell, Michigan, 20 Oct 1967.

Pileus 1.5–3 (5) cm broad, obtuse, the margin at first inrolled, expanding to plane or plano-umbonate, surface thinly viscid, pale dingy vinaceous-cinnamon overall in age, when young the disc "Verona Brown" and the margin paler and with only a few traces of veil remnants along it. Context pallid, odor and taste strongly raphanoid; FeSO$_4$ staining context and stipe olive.

Lamellae very pale dingy vinaceous-cinnamon at first, near "Verona Brown" in age, close, depressed-adnate, moderately broad, edges pallid, not beaded and not stained.

Stipe 6–10 cm long, 4–9 mm thick, equal, splitting readily, solid, becoming tubular, silky to scurfy-fibrillose below, pruinose-scurfy above, *not* discoloring appreciably at the base and when dried concolorous with the pileus margin. Veil fibrillose, very thin, pallid.

Spores 8–10 × 5–5.5 μm smooth under high-dry objective, ellipsoid to ovoid, pale ochraceous in KOH and about the same in Melzer's but finally slightly redder (not dextrinoid).

Hymenium.—Basidia 4-spored, 8–10 μm broad near apex, in mass orange-red in Melzer's fading slowly to orange-ochraceous. Pleurocystidia none. Cheilocystidia fusoid-ventricose and 33–47 × 7–12 μm or cylindric and up to 70 × 5–6 μm, apex obtuse, thin-walled and hyaline in KOH mounts.

Lamellar and pilear tissues.—Gill trama typical for the genus but the subhymenium orange-red in Melzer's. Cuticle of pileus a distinct ixocutis, hyphae 1.5–2.5 μm diam, walls highly refractive, clamps present, no dextrinoid debris evident. Hypodermium hyphoid, with scattered "cells" present, pale clay color in KOH, in Melzer's slightly more reddish brown and dextrinoid debris present. Tramal hyphae typical of the genus, mostly not highly colored in Melzer's or (near the hymenium) yellowish. Clamp connections present.

Habit, habitat, and distribution.—Clustered to gregarious or scattered on sandy soil under spruce, Hell, Michigan, October 20, 1967 (type, MICH).

Observations.—This species is quite similar to *H. pascuense,* hence the species epithet. It differs in the reddish brown pileus when young, the readily splitting stipe, the colored hypoderm (in KOH), the strong raphanoid odor and taste and the vinaceous-cinnamon gills when young.

105. Hebeloma pseudomesophaeum sp. nov. (l)

Pileus 3–5 cm latus, obtusus demum late expansus, hygrophanus, subviscidus, ad marginem fibrillosus vel adpresse squamulosus, fulvus demum cinnamomeus; odor ± pungens; gustus mitis. Lamellae latae, adnatae, confertae, pallide cinnamomeae demum subfulvae. Stipes 3–6 cm longus, 3–7 mm crassus, sursum pallidus, deorsum subfulvescens;

annulus vel fibrillose zonatus, annulus evanescens. Velum pallide argil-laceum, copiosum. Sporae 8–10 (11) × 5–6 μm, ellipsoideae vel ovoi-deae vel phaseoliformes, non dextrinoideae, subleves. Cheilocystidia cylindrica, 36–70 × 4–6 μm vel fusoid-ventricosa, 35–50 × 7–10 × 4–6 μm, vel breviclavata.

Specimen typicum in Herb. Univ. Mich. conservatum est, Smith 88315; legit prope Dexter, Michigan, 12 Oct 1977.

Pileus 3–5 cm broad, obtuse then broadly convex, often irregular in outline, surface viscid, moist, hygrophanous, decorated with patches of pinkish buff to cinnamon-buff fibrils along the margin, disc gla-brous, "Warm Sepia" fading to "Tawny," margin at times ± wood brown or paler, rarely watery zoned or spotted, opaque. Context thin but flesh watery dull cinnamon when mature, fading to pallid buff, odor ± faintly pungent, taste slight; $FeSO_4$ staining base of stipe olive-black, KOH no reaction on cuticle of pileus (or merely brownish).

Lamellae broad, adnate becoming depressed around the stipe, close to crowded, pallid cinnamon becoming subfulvous.

Stipe 3–6 cm long, 3–7 mm thick, equal, often compressed, pallid above when young, soon dingy brown overall but base slowly darkening and the change progressing upward; surface coated with pinkish buff to cinnamon-buff fibrils up to the fibrillose annulus, an-nulus evanescent.

Spores 8–10 (11) × 5–6 μm, pale yellow-brown (clay color) in KOH, not much different in Melzer's (not dextrinoid), smooth (under high-dry objective), ellipsoid to ovoid or in profile some ± bean-shaped.

Hymenium.—Basidia 4-spored, 7–8 μm broad near apex. Pleuro-cystidia none. Cheilocystidia cylindric and 36–70 × 4–6 μm or nar-rowly ventricose and 35–50 × 7–10 × 4–6 μm, and a small number short-clavate, thin-walled and hyaline.

Lamellar and pilear tissues.—Cuticle of pileus a thin but well-defined ixocutis, hyphae 2–4 μm diam, walls gelatinized and clamps present. Hypodermium cellular, of rusty brown cells (in KOH), red-dish brown in Melzer's, not all the pigment incrustations dissolving, no dextrinoid debris observed. Tramal hyphae (both lamellae and pileus) typical for the genus. Clamp connections present.

Habit, habitat, and distribution.—Gregarious in troops of hun-dreds of basidiocarps, under spruce, Dexter, Michigan, October 12, 1977 (type, MICH).

Observations.—The generally fulvous pigmentation, the copious buff-colored fibrillose veil, lack of a distinctive taste, and the watery zones and spots observed on many caps make a distinctive combination of features. The zones and spots are common on species of subgenus

Denudata but very unusual on the veiled species, *H. pseudomeso-phaeum* occurred in great numbers in 1977 and was readily recognized at sight.

106. Hebeloma pumiloides sp. nov.
var. pumiloides (m)

Pileus (1.5) 2.5–4 cm latus, obtusus vel convexus, subviscidus, leviter fibrillosus, glabrescens, ad centrum rufobrunneus, ad marginem pallidus. Contextus fragilis, odor mitis, gustus amarus. Lamellae rufo-cinnamomeae, confertae, angustae, adnatae. Stipes 3–5 cm longus, 3–6 mm crassus, fulvus, fibrillosus vel squamulosus, glabrescens. Velum fibrillosum subochraceum. Sporae 7–9 × 4–5 μm, ellipsoideae vel ovoideae, non dextrinoideae, subleves. Cheilocystidia 29–46 (57) × 4–7 × 4–5 μm, anguste fusoid-ventricosa vel cylindrica.

Specimen typicum in Herb. Univ. Mich. conservatum est, Smith 88232; legit prope Milford, Oakland County, Michigan, 4 Oct 1977.

Pileus (1.5) 2.5–4 cm broad, obtuse to convex, soon dry, margin at first decorated with patches of buff fibrils from the thin fibrillose veil; color "Verona Brown" (rufobrunneus) on disc and whitish ("Til-leul Buff") or slightly darker on the margin (color pattern very nonde-script). Context buffy pallid, odor slight, taste bitter; FeSO₄ staining base of stipe instantly olive-gray; KOH no reaction on pileus cuticle.

Lamellae pale "Verona Brown" (dull pinkish cinnamon), nar-row, adnate, close, not beaded and not spotted.

Stipe 3–5 cm long, 3–6 mm thick, equal, rusty brown in cortex in the base, pallid above, surface fibrillose to squamulose, pallid brown on surface near base at first, becoming dark brown overall in age. Veil fibrillose, thin, pale buff.

Spores 7–9 × 4–5 μm, ellipsoid to ovoid, pale buff in KOH, pale ochraceous in Melzer's, smooth under light microscope, not dextri-noid.

Hymenium.—Basidia 4-spored, 6–7 μm broad near apex, subcy-lindric. Pleurocystidia none. Cheilocystidia 29–46 (57) × 4–7 × 4–5 μm, subcylindric to slightly ventricose near base, apex obtuse, scat-tered along the gill edge, hyaline and thin-walled.

Lamellar and pilear tissues.—Lamellar trama typical of the genus but the subhymenium of ± inflated cells in the region near the pileus trama. Cuticle of pileus a distinct ixocutis the hyphae 1.5–3 μm diam, walls gelatinized, clamp connections present. Hypodermium cellular, dark rusty brown in KOH, patches of encrusting pigmented material on the walls, these vanishing in Melzer's mounts. Tramal hyphae near the hymenium with greatly inflated cells (22–40 μm diam).

Habit, habitat, and distribution.—Gregarious under hardwoods, Oakland County, Michigan, October 4, 1977 (type, MICH).

Observations.—The bitter taste, darkening stipe, and very small spores are a distinctive combination. It differs from *H. pumilum* Lange in the more robust fruiting bodies, the colored veil, and in particular in the shape of the spores in profile. The type variety of *H. pumiloides* has spores 7–9 × 4–5 μm and a buff-colored veil. Both it and var. *sylvestre* were collected in the same area of Highlands Recreation Area on the same day.

107. Hebeloma pumiloides var. sylvestre var. nov. (n)

Pileus 1–3 cm latus, campanulatus vel ± planus, cinereocanescens dein nudus, brunneogriseus demum rufobrunnescens; odor mitis; gustus amarus. Lamellae latae, confertae, adnatae, sordide cinnamomeae. Stipes 1.5–5 cm longus, 2.5–4 mm crassus, deorsum brunnescens, fibrillosus. Velum copiosum, cinereopallidum, evanescens. Sporae 9–11 × 5–6 μm, subellipsoideae, non dextrinoideae. Cheilocystidia 38–46 × 6–9 × 4–6 μm, obtusa; vel subfilmentosa 30–56 × 4–6 μm.

Specimen typicum in Herb. Univ. Mich. conservatum est, Smith 88230; legit prope Milford, Oakland County, Michigan, 4 Oct 1977.

Pileus 1–3 cm broad, conic to convex becoming campanulate to expanded umbonate or plane, surface merely moist when fresh, grayish-hoary but soon naked, color ± "Wood Brown" but developing a "Verona Brown" hue and in age this color overall, opaque at all times. Context white, thick in the umbo, odor slight (not distinctive), taste bitter (distinctly so), $FeSO_4$ staining pileus gray and the stipe olive-gray; KOH on cuticle a dark brown.

Lamellae "Sayal Brown" at maturity, broad, adnate, close, not beaded and not spotted.

Stipe 1.5–4 (5) cm long, 2.5–4 mm thick, equal, no true bulb present, surface grayish-fibrillose from a thin veil, soon naked and unpolished, brownish in the base and pallid above but darkening to dull brown overall.

Spores 9–11 × 5–6 μm, pale clay color in KOH, not much different in Melzer's, smooth to faintly marbled (under high-dry objective); ellipsoid to ovoid, some obscurely inequilateral in profile view, not dextrinoid.

Hymenium.—Basidia 4-spored, 7–8 μm broad near apex. Pleurocystidia none. Cheilocystidia scattered mostly cylindric, 38–46 × 6–9 × 4–6 μm, or ± filamentous and 30–56 × 4–6 μm.

Lamellar and pilear tissues.—Lamellar trama typical of the genus, but diffusely reddish in mounts in Melzer's, color soon fading. Cuticle of pileus a thin layer of gelatinized hyphae (an ixocutis), some with a yellowish brown content as seen mounted in KOH, with pigment-encrusted walls next to the hypodermium. Hypodermium dark rusty brown in KOH with pigment encrusted on the walls, "Hay's Brown" in Melzer's and copious dextrinoid debris forming in the mount in 10–15 minutes. Hyphae of pileus trama typical for the genus. Clamp connections present.

Habit, habitat, and distribution.—Gregarious under hardwoods, in Highlands Recreation Area, Oakland County, Michigan, October 4, 1977 (type, MICH).

Observations.—This taxon differs from var. *pumiloides* in having larger spores and in a ± cinereous to pallid veil. The red flush of the trama in mounts made in Melzer's may be an additional difference.

108. Hebeloma sterlingii (Pk.) Murrill (o)

North Amer. Flora 10: 217. 1917

Inocybe sterlingii Peck, Bull. Torrey Club 33: 217. 1906.

Pileus 1.5–2.5 cm broad, convex to nearly plane, gray or clay colored, center brownish, glabrous, slightly viscid at the center when moist, margin incurved and obscurely fibrillose-appendiculate. Context with a farinaceous taste.

Lamellae adnexed, pallid, becoming cinnamon-brown (from the spores ?).

Stipe 2.5–3.5 cm long, 2–4 mm thick, equal or enlarged near the base, white over exterior, bay-red within, fibrillose from the veil which leaves an evanescent ring (zone), at times the margin of the pileus appendiculate with veil remnants as is the lower portion of the stipe.

Spores 9–12 × 5.5–7 μm, more or less inequilateral in profile view, in face view elliptic to subovate, the wall 0.3 μm thick, surface obscurely rugulose, color as revived in KOH pale yellow, not dextrinoid.

Hymenium.—Basidia 4-spored, 28–32 × 6–7 μm. Pleurocystidia none. Cheilocystidia 32–53 × 4–8 × 3–4 μm, cylindric, usually slightly enlarged near the base (± fusoid-ventricose and the neck ± elongated).

Lamellar and pilear tissues.—Trama of the lamellae of narrow subparallel hyphae. Trama of pileus of radially disposed hyphae. Cuticle of pileus an ixocutis. Hypodermium cellular. Stipe cuticle of repent dry hyphae. Clamp connections widely scattered on the cuticular hyphae of both pileus and stipe.

Habit, habitat, and distribution.—On soil under spruce trees, solitary to gregarious; Trenton, New Jersey, collected by E. B. Sterling, November, 1905 (type, NY). Type studied by Hesler.

Observations.—The features of the spores and the cheilocystidia show that Murrill was correct when he transferred the species from *Inocybe* to *Hebeloma*. The color of the interior of the stipe as described is a very unusual feature in *Hebeloma*. The color of the interior of the stipe at the base may well be a variation of the change to brown found so many times in this subgenus, and, if so, would not be expected to occur in young or freshly maturing specimens.

109. Hebeloma strophosum (Fr.) Sacc. var. strophosum sensu Moser (1978) (p)

Pileus 3–6 cm broad, obtuse to convex, becoming broadly convex to plane, margin inrolled at first, surface glabrous and "Warm Sepia" to "Verona Brown" over disc, margin grayish at first from a coating of fibrils, the edge ± appendiculate, pileus margin at times spotted with patches of veil material but gradually ± glabrescent. Context watery gray to brownish fading to pallid, odor and taste sharp (± raphanoid), $FeSO_4$ staining base of stipe gray; KOH on cuticle of pileus and context not distinctive.

Lamellae pallid ("Tilleul Buff"), soon pale dull brown then ± "Vinaceous-Buff" becoming "Avellaneous," finally near "Sayal Brown" (from the spores ?), adnate, close, finally broad, not beaded and not spotted.

Stipe 4–8 cm long, 5–9 mm thick, equal, pallid to white above the veil-line, slowly avellaneous at base, copiously fibrillose from grayish fibrils, solid, soon becoming hollow; as dried pallid overall including veil fibrils.

Spores 7–9 × 4.5–5 μm, ellipsoid to ovoid, thin-walled and merely yellowish pallid in KOH and Melzer's (not dextrinoid).

Hymenium.—Basidia 4-spored, 7–8 μm wide near apex. Pleurocystidia absent. Cheilocystidia 32–47 × 6–9 μm, neck 4–5 μm diam and apex obtuse to subcapitate; many ± cylindric and some narrowly clavate (5–7 μm near apex), soon becoming somewhat agglutinated.

Lamellar and pilear tissues.—Lamellar trama typical for the genus. Cuticle of pileus an ixocutis, the hyphae 2–5 μm diam and many collapsed, walls refractive, clamps present. Hypodermium cellular, the walls rusty brown or paler (in either KOH or Melzer's reagent), encrusting material not conspicuous, dextrinoid debris present in and near the layer. Tramal hyphae typical of the genus.

Habit, habitat, and distribution.—Gregarious under shrubs and trees of broad-leafed species, Washtenaw County, Michigan, October 1, 1975, Smith 86776.

Observations.—The veil remnants on the stipe and on the margin of the pileus are typically heavy and not significantly colored. The fragments on the pileus are soon obliterated. The species, characteristically, is found under conifers; however, in North America, we find collections of *H. strophosum* intergrading with *H. mesophaeum*. This variable species in its broad sense here in North America, has been found associated with both hardwoods and conifers. For *H. strophosum* we emphasize the heavy veil remnants on the stipe, the ± raphanoid odor and taste, and the relatively small spores.

110. Hebeloma subfastigiatum sp. nov. (q)

Pileus 2–5 cm latus, obtuse conicus demum late convexus vel obtuse umbonatus, ad marginem leviter fibrillosus, pallide argillaceus demum rufobrunneus; odor et sapor leviter raphaninus. Lamellae confertae, latae, adnatae, pallidae demum argillaceae. Stipes 3–8 cm longus, 3–10 mm crassus, sursum pallidus, deorsum demum "Bister" (± spadiceus). Velum pallide argillaceum. Sporae 9–12 × 5.5–7 μm, tarde dextrinoideae, subleves, in laterales subellipsoideae vel obscure inequilaterales. Cheilocystidia 35–54 × 7–10 × 5–6 μm vel 47–63 × 5–7 μm.

Specimen typicum in Herb. Univ. Mich. conservatum est, Smith 67833; legit in terra carbonicola, Washtenaw County, Michigan, 1 May 1964.

Pileus 2–5 cm broad, obtusely conic to convex, expanding to plane or with a low umbo, often remaining convex, near the margin faintly fibrillose-streaked from remnants of the veil, at times with a zone of fibrils on the margin or the latter fringed, surface merely subviscid when fresh, color a pale buff to pinkish buff or on the disc dull cinnamon to reddish brown, often reddish brown at first, in age when water-soaked merely a dingy tawny. Context watery olive-buff; odor and taste slightly raphanoid; $FeSO_4$ no reaction on pileus (base of stipe not tested).

Lamellae close, adnate to adnexed, seceding, moderately broad, equal, edges even, faces pallid then dingy cinnamon to clay color, not beaded and not spotted.

Stipe 3–8 cm long, 3–10 mm thick, equal, pallid above, soon bister from the base upward, brown overall in age, lacerate-fibrillose from remains of the buff-colored veil (the remnants seldom in a zone), apex pallid and silky.

Spores 9–12 × 5.5–7 μm, in Melzer's the surface seen to be faintly marbled, in KOH appearing smooth, pale dingy ochraceous in

KOH, in Melzer's slowly ± reddish tawny; shape ellipsoid to ovoid or in profile tending to be obscurely inequilateral.

Hymenium.—Basidia 4-spored, 7–10 μm broad near apex. Pleurocystidia none. Cheilocystidia fusoid-ventricose to ± cylindric, 35–54 × 7–10 × 5–6 μm for the former, for the latter 47–63 × 5–7 μm, apex obtuse and many with a refractive wall thickening in the interior of the apex.

Lamellar and pilear tissues.—Gill trama typical for the genus but in Melzer's solution the hymenium orange-ochraceous or reddish orange but soon fading. Cuticle of pileus a well-defined ixocutis, the hyphae 1.5–3 μm diam and outer wall ± gelatinized, no dextrinoid debris noted. Clamps present. Hypodermium intermediate in type, ± clay color in KOH, ± reddish brown in Melzer's, incrustations lacking. Tramal hyphae typical for the genus.

Habit, habitat, and distribution.—On a freshly burned-over area, Washtenaw County, Michigan, May 1, 1964 (type, MICH).

Observations.—The habitat, ordinarily, is humus under low moist hardwoods, and the fruiting bodies were produced in an area of about the size of an acre. Hundreds of them were present. The burn was about a month old at the time. We at first classified this material as *H. fastibile*. The veil is buff colored, the spores are a trifle large, and the habitat and time of fruiting make identification with *H. fastibile* doubtful. It differs from *H. pseudofastibile* by its occurrence on a fresh burn in a low hardwood forest lacking any conifers, in the gill edges not being beaded, in having a buff-colored veil, and in the stipe not splitting into longitudinal segments.

111. Hebeloma urbanicola sp. nov. (r)

Pileus 1.5–3 cm latus, obtuse campanulatus, ad marginem fibrillosus, glabrescens, subviscidus, cinnamomeus, tarde ± spadiceus. Contextus pallide brunneus, fragilis; odor et gustus mitis. Lamellae confertae, adnatae demum sinuatae, subdistantes, pallidae dein cinnamomeae. Stipes 2.5–4 cm longus, 2–3.5 mm crassus, albidus, demum brunneus. Velum fibrillosum, pallidum. Sporae 7–9 × 4.5–5 μm, non dextrinoideae, ± ellipsoideae, ± leves, in "KOH" subhyalinae. Basidia tetraspora. Cheilocystidia 26–37 (44) × 6–9 × 3–5 μm, clavata vel subfusoide ventricosa (collum angustum).

Specimen typicum in Herb. Univ. Mich. conservatum est, Smith 34233; legit sub *Piceae*, in Ann Arbor, Michigan, 7 Oct 1949.

Pileus 1.5–3 cm broad, obtusely campanulate to expanded-umbonate, margin incurved and at first fringed with veil remnants, hoary at first from a thin coating of veil fibrils, glabrescent and ± "Sayal

Brown" when fibrils are removed, becoming a pale dull tan as moisture escapes (and on aging and resoaking becoming ± bister). Context thin, brownish, more or less fragile, odor and taste mild.

Lamellae close, becoming adnexed and moderately broad, ± subdistant, pallid at first, dull cinnamon in age and when dried a pale cinnamon (much as in *H. pascuense*), edges not beaded and not stained.

Stipe 2.5–4 cm long, 2–3.5 mm thick, equal, not splitting lengthwise, whitish, in age generally ± discolored, the base merely ± cinnamon as dried. Veil fibrillose, pallid, merely coating the stipe (no annular zones or veil-line showing in dried material).

Spores 7–9 × 4.5–5 μm, very pale both in KOH and Melzer's (not dextrinoid), faintly marbled, thin-walled, oblong, ellipsoid to ovoid.

Hymenium.—Basidia 4-spored, 6–7 μm broad at apex. Pleurocystidia none. Cheilocystidia small, 26–37 (44) × 6–9 × 3–5 μm, clavate and often with a filamentous projection 3–4 μm diam developing from near or on the apex, a secondary septum occasionally noted at the point of origin of the prolongation; many cells simply remaining clavate, all thin-walled hyaline and scarcely refractive.

Lamellar and pilear tissues.—Cuticle of pileus a well-defined ixocutis of narrow (2–3.5 μm) hyphae, clamps present, outer wall gelatinizing or merely refractive. Hypodermium hyphoid or appearing cellular in tangential sections, dull brown in KOH and ± reddish brown in Melzer's, no dextrinoid debris seen. Tramal hyphae typical of the genus. No particular color evident in sections mounted in KOH or in Melzer's; clamp connections present.

Habit, habitat, and distribution.—Gregarious under blue spruce, 1766 Glenwood Rd., Ann Arbor, Michigan, October 7, 1949 (type, MICH).

Observations.—Since the original collection, the odor and taste have been tested repeatedly and the result was always the same. The small cheilocystidia, small spores, lack of odor and taste, lack of color in the veil fibrils, and the pattern of the brownish change in the stipe make a significant combination of characters.

112. Hebeloma velatum (Pk.) Peck (s)

Bull. N.Y. State Mus. 139: 69. 1910

Hebeloma colvini var. *velatum* Peck, Ann. Rep. N.Y. State Mus. 48: 19. 1897.

Pileus 2.5–6 cm broad, at first convex to campanulate-convex or conic, finally expanding to plane or broadly and obtusely umbonate, sometimes the central area depressed, viscid; grayish brown to near "Clay Color" toward the disc which is vinaceous-brown (near "Rood's Brown" to "Cinnamon-Brown"), the edge usually pinkish buff or paler when faded; surface often hoary and with white patches of the veil near the margin or along it and arranged in one or two rows, the veil remnants relatively persistent. Context white, thick in the disc, rather thin on the margin; odor and taste radishlike to nearly mild.

Lamellae narrowly adnate to adnexed, broad, close, at first whitish, finally "Cinnamon-Buff" to "Wood Brown" (grayish brown) or darker, edges white floccose; lamellulae present and usually alternating with lamellae, in ± four ranks.

Stipe 3–6 cm long, 3–8 (11) mm thick, dingy brownish, darker below, pallid near apex and mealy, fibrillose-striate below, equal or tapered toward base, finally hollow, dry. Veil copious, white, fibrillose, leaving an annulus or fibrillose ring near the stipe apex and also leaving patches or zones over the area below the ring.

Spore deposit "Sayal Brown" as air dried. Spores 8–10.5 (11) × 5–6.5 (7) μm, shape in profile view ± inequilateral, in face view subovate to elliptic, wall ± 0.3 μm thick, minutely rugulose (under oil-immersion lens), pale yellowish in Melzer's (not dextrinoid), paler in KOH.

Hymenium.—Basidia 4-spored, 7–9 μm broad near apex. Pleurocystidia none. Cheilocystidia 30–40 (70) × 5–9 (13) μm, subcylindric above a ± ventricose region near the base.

Lamellar and pilear tissues.—Lamellar trama typical of the genus. Cuticle of pileus a well-defined ixocutis, or in age an ixolattice. Hypodermium cellular, the walls dark brown in KOH, some hyphae in the cuticular area encrusted. Clamps present.

Habit, habitat, and distribution.—Solitary to gregarious or caespitose under cottonwood, summer and fall.

Observations.—Peck's type has been studied, but the species still presents problems in relation to *H. strophosum,* and *H. mesophaeum* and its variants. The ± inequilateral spores in profile view distinguish it from the latter. We might point out, however, that for a time we carried it in a preliminary manuscripts as a variant of *H. mesophaeum.* Var. *castaneum* of the latter, for instance, differs mainly in having a thin veil and in being associated with conifers.

MATERIAL CITED

Hebeloma affine: S-42918(type)
        agglutinatum: S-88328(type)
        aggregatum: S-90057(type)
        alpinicola: S-58632(type)
        amarellum: S-9339(type)
        angelesiense: S-17096(type); S-17184
        angustifolium: S-88295(type)
        aurantiellum: S-55477(type)
        barrowsii: Barrows-3056(type)
        bicoloratum var. bicoloratum: S-88316(type); S-88323; S-88325; S-88326
                     coloradense: S-89594; DBG-7956
        boulderense: DBG-7959(type)
        brunneodiscum: S-86872(type)
        brunneomaculatum: W-K-7(type)
        chapmanae: DBG-2473(type); Barrows-2021; Thiers-20774
        cinereostipes: S-88733(type); S-90158
        coniferarum: S-86924(type); S-90060?
        corrugatum: S-89266(type); S-87089; S-89078; S-89107; S-89108; S-89267;
                     S-89889; S-90213
        dissiliens: S-89723(type)
        evensoniae: DBG-7948; DBG-7960
        farinaceum: Murrill-126(type)
        fastibile: Kauffman's Wyoming collection; no type
        felleum: S-90628
        flaccidum: S-90452(type)
        fragrans var. fragrans: S-90594(type); S-90620; S-90629; S-90630
                     intermedium: S-90636(type)
        fuscostipes: S-90501(type)
        glabrescens: S-89053(type)
        gregarium: Peck's type
        griseocanescens: S-88697(type)
        griseocanum: S-90121(type)
        griseovelatum: S-86929(type)
        hesleri: S-86922(type)
        hydrocybeoides: S-36228(type)
        idahoense: S-66049(type); S-53293; S-86922; S-90152; S-90367
        immutabile: S-17538(type)
        indecisum: S-89584(type)
        insigne: S-89184(type); S-5245; S-87013; S-89461; S-90195; S-90197
        juneauense: W-K-5958(type)
        kanousiae: Kanouse August 28, 1923(type)
        kauffmanii: Kauff.-9-7-23(type)
        kelloggense: S-73592(type)
        kemptonae: W-K-5143(type)
        kuehneri: S-88981; S-88982

Hebeloma latisporum: S-55278(type); S-88817
     limacinum: S-89110(type)
     littenii: S-10-29-80(type)
     luteobrunneum: S-88819(type)
     lutescentipes: S-24018(type)
     marginatulum var. fallax: S-89724(type); S-90276
              marginatulum: S-53299(type); S-87552; S-88911
          var. prov. proximum: S-90631
     maritinum: S-9008(type)
     mesophaeum var. aspenicola: S-90192(type); S-90194
             bifurcatum: S-88816(type)
             castaneum: S-88922(type); S-87445
             duplicatum: S-90058A(type)
             fluviatile: S-89471(type); S-89485; S-90008
             imitatum: S-53134(type); S-89567; S-90198
             insipidum: S-88791(type); S-89723; Solheim-6235
             lateritium: Murrill-295(type)
             longipes: S-64681(type)
             mesophaeum: S-17242; S-49417; S-88327; S-90210
             similissimum: S-87091(type)
             subobscurum: S-46587(type)
             velovinaceum: S-51374(type); S-52226
     naucorioides: S-32-580(type)
     nigromaculatum: S-19314(type); S-89678
     obscurum: S-88923(type)
     occidentale: S-19106(type)
     ollaliense: S-23737(type); S-23715; S-90160
     olympianum: S-17972(type); S-71130
     oregonense var. atrobrunneum: S-89268(type)
              oregonense: S-24289(type); S-24290; S-51737; S-90632
     pallescens: S-89008(type)
     pallido-argillaceum: S-90600(type)
     parcivelum: S-23703(type)
     pascuense: Peck's type
     perfarinaceum: S-90421(type)
     perigoense: DBG-4877(type); S-44396; S-44427; S-45243
     perplexum: S-90498(type)
     piceicola: S-89135(type); S-89733A
     pinetorum: S-24799(type)
     pitkinense: S-86865(type)
     praecaespitosum: S-89627(type); S-90477; S-90916
     praelatifolium: S-89109(type); S-73961; S-87486
     praeolidum: S-17169(type); S-17208; S-17263; S-17607; S-17702; S-17703;
            S-17704; S-19714; S-24954; S-27870; S-51864; S-55626; S-70460
     proximum: S-75218(type)
     pseudofastibile var.distans: S-68874(type); DBG-7955
             pseudofastibile: S-90425(type)
     pseudomesophaeum: S-88315(type)
     pseudostrophosum: S-14443(type); S-44533; S-19967; S-27388; S-76618;
            S-90916; Ammirati-6027; Barrows-1190

Hebeloma pumiloides: var. pumiloides: S-88232(type)
                        sylvestre: S-88230(type)
            pungens: S-24293(type)
            remyi: S-87045; S-89846; S-89850; S-89888; S-90434
            repandum: S-86910(type)
            riparium: S-88621(type); S-48600; S-49261
            salmonense: S-70148(type)
            sanjuanense: S-51736(type)
            solheimii: S-89471(type); Solheim-6235
            stanleyense: S-46221(type); S-70148; S-90420
            sterlingii: Peck's type
            strophosum var. occidentale: S-44190(type); Kauffman-Wyoming
                        strophosum: S-86776; S-6072
            subannulatum: S-89095(type); Kauff-8-26-20(Colorado)
            subargillaceum: DBG-7947(type)
            subboreale: S-89006(type); S-87408
            subcapitatum: S-89093(type); S-89234
            subfastigiatum: S-67833(type)
            subhepaticum: S-18029(type); S-56753; S-56757; S-56838; S-57020; S-57047
            sublamellatum: S-87044(type)
            subrimosum: S-90148(type)
            subrubescens: S-90090(type)
            subsacchariolens: S-90215
            substrophosum: S-87043(type)
            subumbrinum: S-88979(type)
            subviolaceum: S-88843(type)
            testaceum: S-90056
            trinidadense: S-57047(type)
            urbanicola: S-34233(type)
            utahense: McKnight-F-1448(type)
            velatum: Peck's type
            vinaceogriseum: S-46587(type)
            vinaceoumbrinum: S-89950(type)
            wellsiae: W-K-16(type)
            wells-kemptonae: W-K-5304(type)

LITERATURE CITED

Arora, David. 1979. Mushrooms Demystified. Ten Speed Press, Berkeley, California. 668 pp.

Bohus, G. 1972. *Hebeloma* Studies I. Ann. Hist-Nat. Mus. Nationales Hungariae 64:71–78.

Bruchet, G. 1970. Contribution a l'étude du genre *Hebeloma* (Fr.) Kummer. Partie Spécial. Bull. Soc. Linn. de Lyon, 39 année, supp. 6. 131 pp.

Fries, Elias M. 1821. Systema Mycologicum. Vol. 1. Upsaliae. 374 pp.

———. 1836–38. Epicrisis Systematis Mycologici. Upsaliae. 610 pp.

———. 1863. Monographia Hymenomycetum Sueciae. Vol. 2. Upsaliae. 355 pp.

Guzmán, Gastón. 1980. Three new sections in the genus *Naematoloma,* and description of a new tropical species. Mycotaxon 12:235–40.

Hacskaylo, E. , and G. Bruchet. 1972. Hebeloma as mycorrhizal fungi. Bull. Torrey Bot. Club. 99:17–20.

Hesler, L. R., and Alexander H. Smith. 1963. North American Species of *Hygrophorus.* University of Tennessee Press, Knoxville. 416 pp.

———. 1979. North American Species of *Lactarius.* University of Michigan Press, Ann Arbor. 841 pp., 154 pls.

Horak, E. 1968. Synopsis generum Agaricalium (Die Gattungstypen der Agaricales). Beiträge zur Kryptogamenflora der Schweiz. Vol. 13. 741 pp.

Kauffman, Calvin H. 1918. The Agaricaceae of Michigan. Vol. I. Mich. Geol. and Biol. Surv. Pub. 26, Biol. Ser. 5. Lansing. 924 pp.

Kummer, Paul. 1871. Der Führer in die Pilzkunde. Zerbst.

Lange, Jakob E. 1938–40. Flora Agaricina Danica III–V. Soc. Advan. Mycol. in Denmark and Danish Bot. Soc., Copenhagen. 5 vols.

Lincoff, Gary, and D. H. Mitchel, M.D. 1977. Toxic and hallucinogenic mushroom poisoning. Van Nostrand-Rheinhold Co., New York and London. 267 pp., 28 pls.

Miller, Orson K. 1972. Mushrooms of North America. E. P. Dutton Co., New York. 359 pp., 292 color figs.

Mitchel, D. H., S. W. Chapman, and M. L. Farr. 1980. Notes on Colorado fungi IV: Myxomycetes. Mycotaxon 10:299–349.

Moser, Meinhard. 1978. Die Röhrlinge und Blätterpilze (Agaricales). In Helmut Gam's Kleine Kryptogamen Flora Bd. IIb/2. Gustav Fisher, Stuttgart. 532 pp.

Murrill, W. A. 1917. *Hebeloma* (Fries) Quélet. In North Amer. Flora 10:215–27.

Peck, Charles Horton. 1896. Rep. N.Y. State Bot. 1895. Ann. Rep. N.Y. State Mus. 49:5–69. Albany, New York.

———. 1897. Rep. N.Y. State Bot. 1896. Ann. Rep. N.Y. State Mus. 48:5–114. 2d. ed. Albany, New York.

———. 1900. Rep. N.Y. State Bot. 1899. Ann. Rep. N.Y. State Mus. 53:821–67. Albany, New York.

———. 1910. Rep. N.Y. State Bot. 1909. Bull. N.Y. State Mus. 139:1–114. Albany, New York.

Pomerleau, René. 1980. Flore des Champignons au Québec. Montreal. 652 pp., color pls. i–xlviii.

Ricken, Adalbert. 1915. Die Blätterpilze (Agaricaceae) Deutschlands und der angrenzenden Länder. Leipzig. 480 pp.

Ridgway, Robert. 1912. Color Standards and Color Nomenclature. Published by the author. Washington, D.C. 44 pp., 53 pls.

Romagnesi, Henri. 1965. Études sur le genre *Hebeloma*. Bull. Tr. Soc. Myc. Fr. 81:321–44.

Singer, R. 1961. Type studies on Basidiomycetes-X. Personia 2:1–62.

———. 1962. The Agaricales in Modern Taxonomy. J. Cramer, Weinheim. 915 pp.

———. 1975. The Agaricales in Modern Taxonomy. 3d rev. ed. J. Cramer, Weinheim. 912 pp.

Smith, Alexander H. 1947. North American Species of *Mycena*. University of Michigan Press, Ann Arbor. 521 pp., 99 pls.

———. 1972. The North American species of *Psathyrella*. Mem. New York Bot. Garden 24:1–633. 95 pls.

Smith, Alexander H., and L. R. Hesler. 1968. The North American species of *Pholiota*. Hafner Pub. Co., New York and London. 402 pp., 115 pls.

———, and Rolf Singer. 1964. A monograph on the genus *Galerina* Earle. Hafner Pub. Co., New York and London. 384 pp., 20 pls.

———, and Nancy S. Weber. 1980. The Mushroom Hunter's Field Guide. Rev. ed. University of Michigan Press, Ann Arbor. 316 pp., 282 color figs.

ILLUSTRATIONS

TEXT FIGURES

The magnification in the figures as reproduced are: for the spores
660 ×, and for the cystidia ± 370 ×.

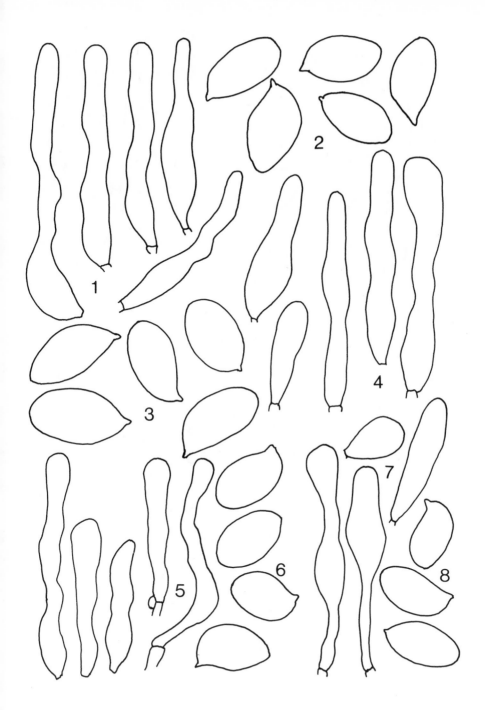

Figs. 1–8. *H. subrimosum:* fig. 1, five cheilocystidia; fig. 2, five spores. *H. aggregatum:* fig. 3, five spores; fig. 4, five cheilocystidia. *H. utahense:* fig. 5, five cheilocystidia; fig. 6, four spores. *H. pseudofastibile* var. *distans:* fig. 7, four cheilocystidia, fig. 8, three spores.

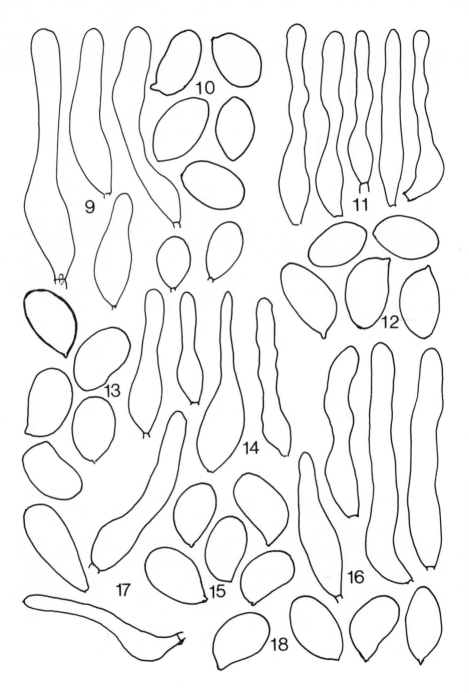

Figs. 9–18. *H. mesophaeum* var. *aspenicola:* fig. 9, six cheilocystidia; fig. 10, five spores. *H. vinaceoumbrinum:* fig. 11, five cheilocystidia; fig. 12, five spores. *H. subviolaceum:* fig. 13, five spores; fig. 14, four cheilocystidia. *H. nigromaculatum:* fig. 15, five spores; fig. 16, four cheilocystidia. *H. substrophosum:* fig. 17, three cheilocystidia; fig. 18, four spores.

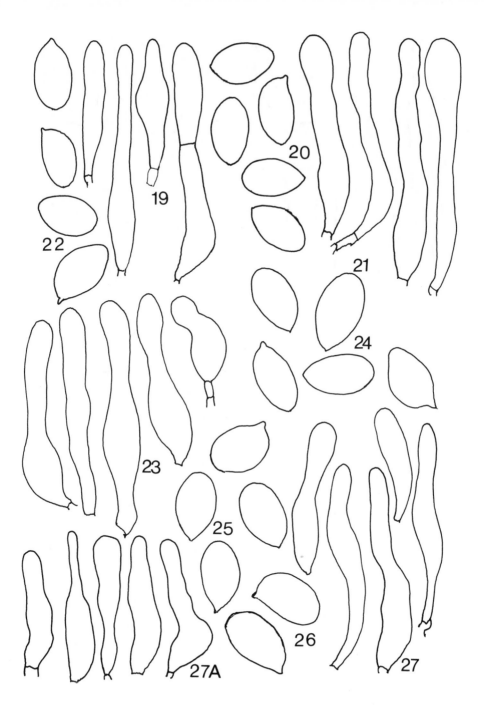

Figs. 19–27A. *H. angelesiense:* fig. 19, four cheilocystidia; fig. 22, four spores. *H. pseudostrophosum:* fig. 20, five spores; fig. 21, four cheilocystidia. *H. bicoloratum* var. *coloradense:* fig. 23, five cheilocystidia; fig. 24, five spores. *H. alpinicola:* figs. 25 and 26, six spores; fig. 27, five cheilocystidia. *H. brunneomaculatum:* fig. 27A, five cheilocystidia.

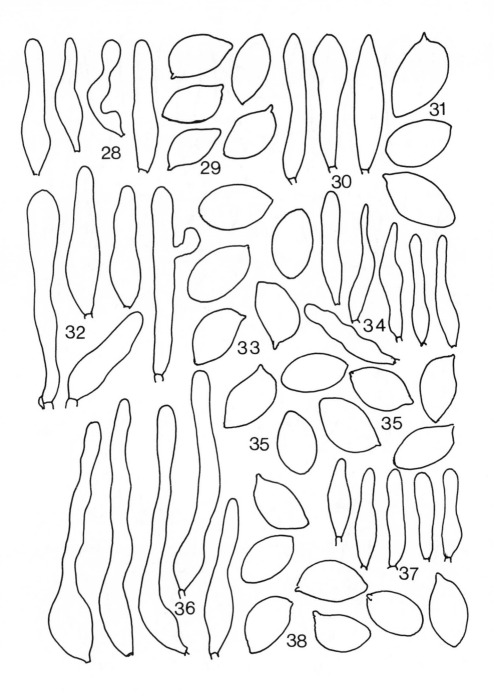

Figs. 28–38. *H. subhepaticum:* fig. 28, four cheilocystidia; fig. 29, five spores.
H. parcivelum: fig. 30, three cheilocystidia; fig. 31, three spores. *H. praecaespitosum:*
fig. 32, five cheilocystidia (from aberrant fruiting body); fig. 33, five spores (from aberrant fruiting body). *H. amarellum:* fig. 34, six cheilocystidia; fig. 35, seven spores. *H.
maritinum:* fig. 36, five cheilocystidia; fig. 37, five cheilocystidia. *H. aurantiellum:* fig.
38, seven spores.

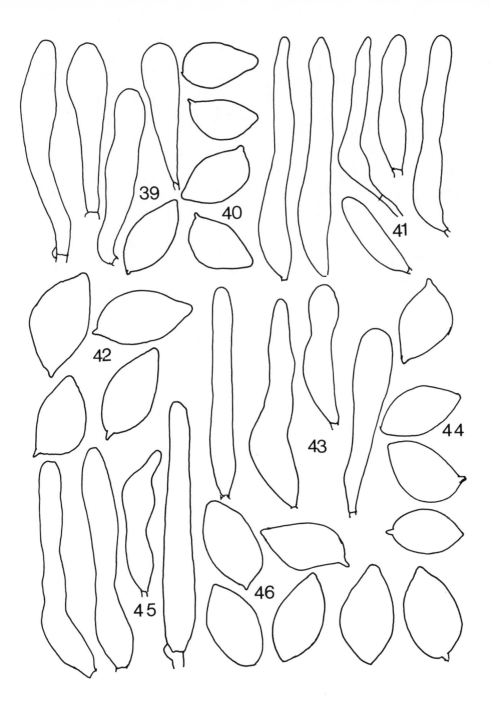

Figs. 39–46. *H. salmonense:* fig. 39, four cheilocystidia; fig. 40, five spores. *H. piceicola:* fig. 41, six cheilocystidia; fig. 42, four spores. *H. oregonense:* fig. 43, four cheilocystidia; fig. 44, four spores. *H. marginatulum* var. *fallax:* fig. 45, four cheilocystidia; fig. 46, six spores.

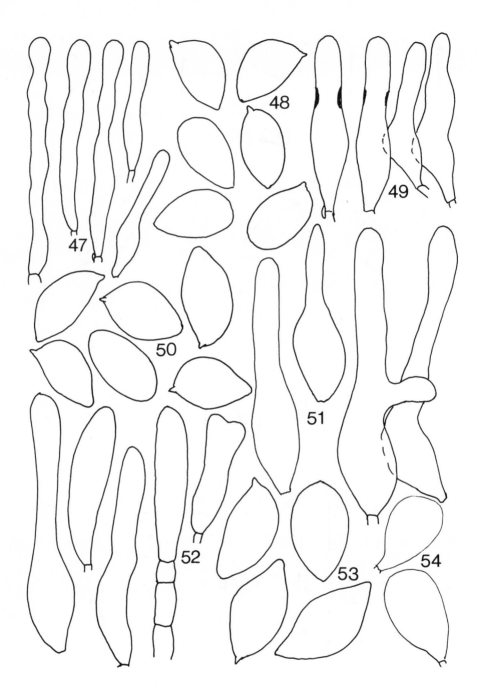

Figs. 47–54. *H. indecisum:* fig. 47, five cheilocystidia; fig. 48, six spores. *H. subrubescens:* fig. 49, four cheilocystidia; fig. 50, six spores. *H. praecaespitosum:* fig. 51, four cheilocystidia; fig. 53, four spores (typical size); fig. 54, two saccate cheilocystidia. *H. occidentale:* fig. 52, five cheilocystidia.

PLATES

PLATE 1

S-17096

Hebeloma angelesiense   × 1

PLATE 2

S-14443

Hebeloma pseudostrophosum   × 1

PLATE 3

S-87043

Hebeloma substrophosum × 1

PLATE 4

Hebeloma praeolidum   × 1

S-17208

PLATE 5

A. Hebeloma praecaespitosum   × ½   S-89627

B. Hebeloma aggregatum   × 1   S-90057

PLATE 6

A. Hebeloma piceicola   × 1   S-89135

B. Hebeloma remyi   × 1   S-89846

PLATE 7

S-90197

Hebeloma insigne   × 1

PLATE 8

Hebeloma naucorioides   × 1

S-32-580

Plates 9–14 showing spore surface configuration as illustrated with the aid of the scanning electron microscope. As can be seen in the illustrations, the details of the spore ornamentation, as seen with aid of the SEM (on the species studied to date), show a pattern of low profile.

An oversimplistic explanation of the formation of the ornamentation of the spores, on the species illustrated, suggests that the ornamentation is produced by a thin more or less viscous "ectoperisporial layer" which becomes pulled apart into the pattern shown, or the layer is gradually eroded in part by enzyme action. Further studies, obviously, are necessary.

PLATE 9

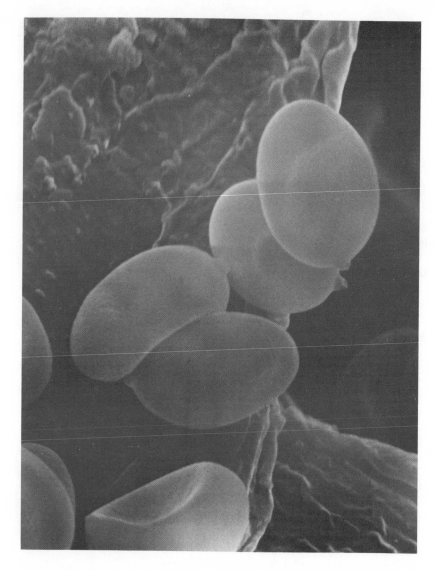

Hebeloma evensoniae showing an essentially smooth spore, DBG-7960.

PLATE 10

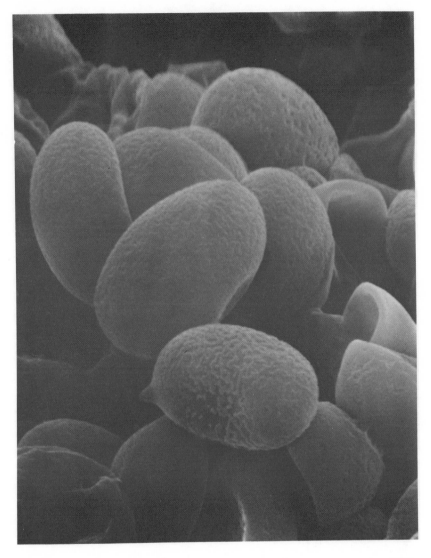

Hebeloma bicoloratum var. *coloradense* showing a fine obscure ornamentation, DBG-7956.

PLATE 11

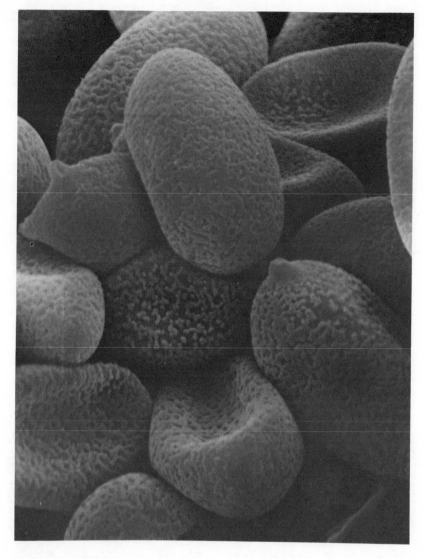

Hebeloma pseudofastibile var. *distans,* showing the way the "ectoperisporial layer" forms obscure but distinctive patterns as it breaks up (or becomes eroded away), DBG-7955.

PLATE 12

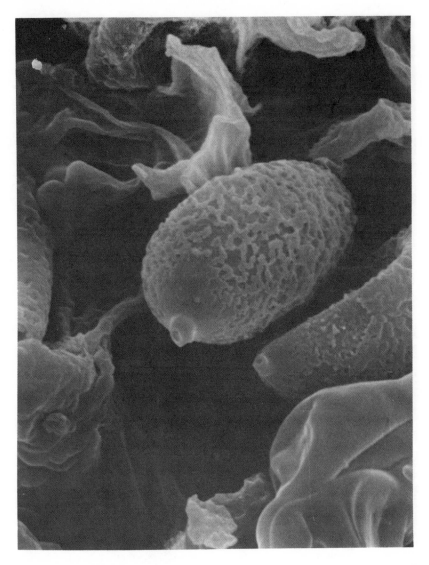

Hebeloma boulderense showing a smooth area remaining at the base of the spore with the rest of the surface ornamented DBG-7959.

PLATE 13

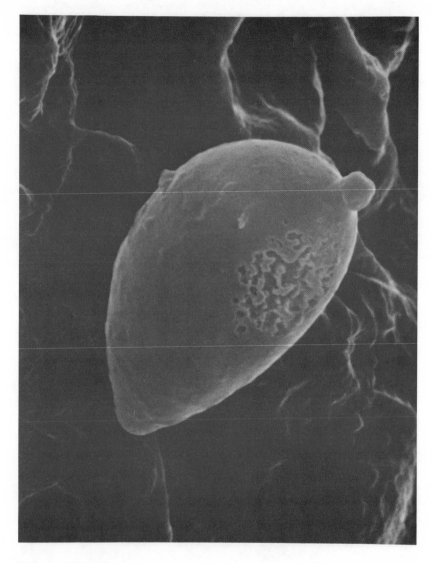

Collection DBG-7957 (ornamentation apparently beginning to form).

PLATE 14

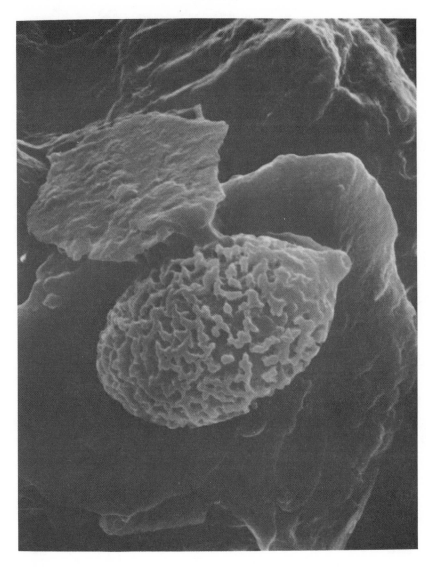

Collection DBG-7957 (same as above but a mature spore with well-developed ornamentation).

INDEX

Page numbers in boldface refer to page on which the description occurs.